YOU ARE NOT STUPID
YOU WERE TARGETED

Reclaim Control of Your Mind, Emotions, and Decisions in an Era of AI, Manipulation, and Engineered Deception

YOU ARE NOT STUPID YOU WERE TARGETED

DR. ALDRIC MARSHALL

Award Winning Author

You Are Not Stupid You Were Targeted

Cover Designer: M.A. Rehman

Interior Formatting: Zahidul

www.global3technology.com

DISCLAIMER

This book is intended solely for informational and educational purposes. While every effort has been made to ensure the accuracy and reliability of the information presented, the author makes no representations or warranties regarding the completeness, accuracy, or applicability of the content.

The material in this book addresses topics related to artificial intelligence, manipulation, influence, and decision-making. It is intended to raise awareness and promote critical thinking, not to provide legal, psychological, financial, or professional advice. Readers are encouraged to exercise their own judgment and seek appropriate professional guidance when necessary.

LIMITATION OF LIABILITY

Under no circumstances shall the author be liable for any loss, harm, or damage, direct or indirect, resulting from the use or misuse of the information contained in this book.

The reader assumes full responsibility for their actions, decisions, and interpretations.

PURPOSE STATEMENT

This book was written to increase awareness of how modern influence operates, particularly in an era when artificial intelligence can shape perception, urgency, and trust in ways that are not always visible.

It is not intended to create fear, but to promote clarity, discipline, and informed decision-making.

Published by:

Global 3 Technology & Intelligence (G3TI)

Author: Dr. Aldric Marshall

Award-Winning Author

ISBN: 979-8-9963484-0-4

DEDICATION

To everyone who thought, "I should have known better."

To those who stayed silent… who bore the weight of a moment no one else saw… who replayed it again and again…

This is for you.

You are not stupid. You were targeted.

ACKNOWLEDGEMENTS

I would like to express my sincere appreciation for those whose work, leadership, and influence continue to shape the evolving conversation about technology, security, and human awareness.

Satya Nadella, your leadership at Microsoft and your emphasis on responsible AI have been critical in shaping how technology can serve humanity while upholding ethical boundaries. Your work continues to influence how we think about innovation and accountability.

Sundar Pichai, your stewardship of Google and its global information ecosystem has redefined how billions of people access and interpret knowledge. Your commitment to advancing AI responsibly remains essential in a world where information can both empower and shape.

Sam Altman, your leadership in advancing artificial intelligence has accelerated global awareness of its potential and risks. Your work has helped bring critical conversations about AI safety and human impact to the forefront.

Dr. Geoffrey Hinton, your pioneering work in artificial intelligence has helped shape the very systems that now influence how information is created, interpreted, and trusted. As the world begins to understand both the power and the risks of these technologies, your voice continues to underscore the importance of awareness, responsibility, and the human impact of intelligent systems.

Jen Easterly, your leadership in cybersecurity and national resilience reflects the growing reality that modern threats extend beyond systems to human perception and decision-making. As influence becomes more sophisticated and less visible, your work continues to underscore the importance of awareness, preparedness, and proactive defense.

Tim Cook, your commitment to privacy and user protection, has set a standard in an era when personal data and digital identity are constantly at risk. Your stance on protecting individuals in the digital age is more relevant than ever.

Mark Zuckerberg, your platforms have fundamentally changed how people connect, communicate, and consume information. With that reach comes a unique responsibility to address how influence and perception are shaped in real time.

Dr. Shoshana Zuboff, your work has brought critical attention to how data and digital systems are used to shape behavior, perception, and decision-making. As influence becomes more embedded in everyday interactions, your insights continue to underscore the importance of awareness and understanding in a rapidly evolving digital world.

Dr. Fei-Fei Li, your advocacy for human-centered artificial intelligence is a powerful reminder that technology must be developed with humanity at its core. Your contributions to ethical AI continue to guide the course of innovation.

This acknowledgment is not simply about recognition; it is about alignment.

The work presented in this book is part of a larger global conversation, one that requires collaboration among technology leaders, security professionals, and everyday individuals.

As artificial intelligence continues to evolve, so does the responsibility to ensure that people are not only protected but also informed.

The G3TI Defensive Intelligence Manifesto

We live in a time when influence no longer announces itself. It does not arrive with force or warning. It arrives quietly, through messages that feel familiar, voices that sound trusted, and moments that feel urgent enough to bypass reflection. What once required effort and proximity can now be executed at scale, refined by data, accelerated by speed, and powered by artificial intelligence. The greatest shift is not only in what technology can do but also in how easily human perception can be guided before awareness has time to respond.

This manifesto is a declaration of clarity in that environment.

We reject the idea that those who are influenced are unintelligent. We reject the assumption that a single misjudgment defines a person's capacity or character. We reject the silence that follows manipulation, the quiet weight of "I should have known." People are not failing to think. They are being moved before thinking has the chance to fully engage. The landscape has changed. The methods have evolved. The pace has accelerated. Yet the human response has not received the same level of attention.

We name what is happening without exaggeration or denial. Influence is engineered. Urgency is manufactured. Trust is simulated. Decisions are shaped in real time by systems designed to anticipate behavior and guide outcomes. The point of failure is not at the end of an event. It is at the beginning, at the moment perception is formed and pressure is applied. By the time a person questions what has happened, the decision has already been made.

We refuse to operate in that space unconsciously.

We choose to become aware of the moment before acting. We choose to recognize pressure when it appears, even when it feels reasonable, even when it sounds familiar, even when it arrives through channels we trust. We choose to slow down in a world that rewards speed. We choose to verify in a culture that normalizes assumptions. We choose to think clearly when the environment is designed to trigger a reaction.

This is not hesitation. This is discipline.

We do not reject technology. We do not retreat from progress. We do not respond with fear. We understand that innovation carries responsibility and that awareness must evolve alongside capability. We believe artificial intelligence can advance human potential, but we also recognize that without awareness, it can influence human behavior in ways that are not immediately apparent. Our position is not against progress. It is for clarity within progress.

We accept that protection is no longer limited to physical security or system integrity. Protection now includes the mind, the emotional response, and the decision itself. To protect a person is to protect the moment of choice. When that moment is compromised, everything that follows is affected.

When that moment is preserved, control remains intact.

We refuse to let shame silence experience. What has happened to one can happen to many. What has been learned must be shared. Awareness is not built through theory alone but through truth that is acknowledged and understood. We speak openly so others can recognize what they have felt but could not explain. We replace judgment with understanding and isolation with clarity.

We operate with intention. We do not move because something feels urgent. We do not trust because something seems familiar. We do not decide because pressure demands it. We paused. We observe. We question. We confirm. We act only when we are certain the decision is truly ours.

This is the standard.

We understand that leadership now requires more than direction. It requires awareness of how influence operates at scale and how quickly it can shape behavior. Those who build systems must consider the human experience within them. Those who protect themselves must consider the unseen moments when harm begins. Those who communicate must recognize the weight of trust and how easily it can be replicated.

We stand at the intersection of technology and human behavior, holding a clear position: progress must not outpace understanding.

Global 3 Technology & Intelligence exists to support that position. It exists to raise awareness to the point where influence begins, to strengthen an individual's ability to recognize it, and to help communities, institutions, and leaders respond to a world in which the most significant threats are often the least visible. This is not a reactionary effort. It is a proactive discipline. It is a commitment to thinking clearly in environments designed to cloud judgment.

We are not passive observers of this shift. We are participants in shaping how people respond to it.

We understand what has been missing, not because it was invisible, but because it was designed to be overlooked. We recognize that decisions can be shaped without force, that trust can be built without a relationship, and that reality can be presented in ways that feel unquestionable in the moment. We accept that this is the environment in which we operate, and we choose to meet it with awareness rather than assumption.

We do not move blindly. We do not react automatically. We do not surrender our judgment.

We take control of the moment before making the decision. We hold that line.

We are not stupid.

We were targeted.

And now we are aware.

"The moment you understand the game; you stop being the piece."

CONTENTS

Preface

Why This Book Had to Be Written

There is a version of this book that could have been written purely from research. From data. From reports. From analysis. From observation. But this is not that book.

This book was born out of experience. Not once. Not twice. But on several occasions, I found myself in situations where I had been misled, manipulated, or caught off guard in ways I should have seen, but didn't.

And each time, the same questions echoed in my mind: How did this happen? Why didn't I see it? What did I miss? And more importantly, how many other people are experiencing this right now and aren't saying anything?

Because that's the part no one talks about. The silence. The embarrassment. The shame. The quiet reply of the moment. The decision not to tell anyone. The internal thought: "I should have known better."

I understand that feeling. I also recognize how risky silence can be. When people stay quiet, they don't learn. Without learning, they don't adapt, without adapting, they stay vulnerable.

That is where I found myself at a crossroads. I had two choices: I could stay an uninformed victim, remain silent, carry the shame, and hope it wouldn't happen again... Or I could stand my ground, learn everything I could, understand how manipulation truly works, study the patterns, build systems, and most importantly... take action on a large scale.

I chose the second path. That decision led to the creation of Global 3 Technology & Intelligence, not just as a company but as a mission. A mission to protect people, expose manipulation, build intelligent systems, and educate individuals, families, communities, and institutions. It aims to ensure that what happened to me, and what happens to millions of people every day, does not go unchecked.

Because this is bigger than just individual moments. It's about a world that is changing, where deception is growing more advanced, manipulation more precisely, pressure more engineered, and people are more easily influenced without realizing it.

This book is my response to that reality. It is not based on theory alone. It is written from experience, observation, discipline, and a refusal to stay silent. If you're reading this, understand this clearly: This isn't a book about making you fearful. It's about making you aware, disciplined, and in control. Because once you understand how these systems work... you'll never see the world the same way again. And that's not a loss, it's power.

Introduction

The World Has Changed, But Most People Haven't

There was a time when most people believed they could trust what they saw. A message meant what it said. A voice belonged to the person speaking. A request came from the person it appeared to come from and if something felt urgent, it was probably real, that time has changed.

We now live in a world where a message can be made to seem perfect yet be completely false, a voice can sound familiar yet not belong to the person you think it does, urgency can be manufactured on demand, trust can be artificially built, and people can be moved to action without ever realizing they were influenced.

This is not science fiction. It happens every day, to professionals, families, elders, young people, businesses, law enforcement, and entire communities. And the most dangerous part is this: most people still act as if the world had not changed.

They still trust based on appearance, respond emotionally, act on urgency, and assume based on familiarity.

And that gap between how the world actually works now and how people still think it works is where manipulation occurs.

THIS BOOK CLOSES THAT GAP

This book isn't about making you suspicious of everything. It's about helping you recognize what truly matters. It's about assisting you:

- identify pressure before it takes over you.
- understand manipulation before it affects you.
- slow down before making a decision you can't undo.
- verify before trusting.
- operate clearly in situations designed to confuse you

THIS IS NOT ABOUT BEING PARANOID

Let's be clear. You don't need to live in fear. You don't have to question everything endlessly. You don't have to lose your ability to trust others.

That is not the objective. The aim is balance.

DISCIPLINED TRUST

The kind of trust that: doesn't rely on emotion alone, doesn't fall apart under pressure, doesn't make assumptions without verification, and doesn't relinquish control too easily.

A NOTE ON HOW THIS BOOK IS WRITTEN

As you progress through this book, you may notice intentional choices. Some sections will seem quick and simple, almost like summaries. Other sections will explore more deeply, being detailed, layered, and thorough. That is not an inconsistency. That is intentional design.

WHY THIS MATTERS

In real life, manipulation doesn't come in just one form. Sometimes it happens quickly, and you need something you can remember easily. Other times, it unfolds slowly, and you need a deeper understanding to spot it.

THIS BOOK IS BUILT TO MATCH REALITY

The concise sections are designed to help you recall key principles under pressure, remember what matters in real time, and avoid being over-whelmed when clarity is most needed.

The deeper sections are there to help you fully understand the patterns, see the structure behind manipulation, and build long-term awareness and discipline.

THIS IS STRATEGIC, NOT ACCIDENTAL

If every page were too dense, you would burn out. If every page were too light, you would lack depth.

This book is designed to provide you with clarity in the moment and long-term understanding.

Think Of It Like This

Some pages are designed for quick review, while others require in-depth study.

And Here Is The Truth

In high-pressure situations, you won't have time to remember everything, but you'll remember something. Sometimes, that one thing is enough to prevent a bad decision.

This Book Is A System

Every chapter you are about to read connects to the next. You will learn:

- how manipulation really works
- why do smart people still fall for it?
- how pressure is designed
- how to recognize the moment before a mistake occurs
- how to verify truth in a world where things can be faked.
- how to become a human firewall
- how to safeguard your family
- how institutions and communities are impacted
- how law enforcement and intelligence must evolve
- what to do after harm has already occurred
- how to rebuild your mindset
- how to operate at a higher level every day
- and ultimately… how to maintain trust in a changing world.

This Is Personal

This isn't just information to me; it's something I had to learn the hard way. Honestly, that's exactly why this book stands out. Because I'm not writing from a distance. I'm writing from a place that understands what it feels like to miss something, to replay it, to question yourself, and to rebuild stronger.

This Is Your Turning Point

As you go through this book, you'll start to notice something: Situations that once felt normal... will begin to feel different. Messages you'd have responded to... you'll now pause before responding. Requests that once would have moved you... you'll question. Pressure that used to control you... will lose its hold.

That is not fear. That is consciousness.

Final Word Before You Start

This book isn't here to make you perfect. It's here to prepare you.

Because the goal isn't to eliminate every possible risk; it's to make sure that when the moment comes, and it will, you're not the same person you were before.

Introduction Takeaway

You don't need to know everything, but you do need to learn to slow down, think clearly, and safeguard your decisions in a world built to rush you.

PART I

THE NEW BATTLESPACE

CHAPTER 1

The War Has Moved in to The Mind

There was a time when danger was clear. You could see it coming down the street, hear it from a distance, and feel it in the room.

Today, danger arrives in a text message. It sounds like a familiar voice. It feels like someone who understands you. And that is the problem.

We are living in an era when deception has become smarter, more scalable, and more precise. The battleground has shifted from physical confrontation to psychological influence. What once required proximity, effort, and risk can now be done anywhere in the world, with speed and automation. The modern attack is no longer about breaking into your house. It's about infiltrating your mind.

The New Reality: Attacks Without Contact

You don't have to be in the same city. You don't have to be in the same country. You don't even have to be a real person.

With the rise of artificial intelligence, synthetic media, and automated communication systems, people can now receive messages that appear personal, authentic, and urgent, even when no real relationship exists.

A voice can be cloned. A face can be generated. A conversation can be simulated. And the person on the other end is expected to decide in real time. That's the point where everything happens.

The True Target Is Not Your Money, It Is Your Judgment

Most people think scams are about money, but they are not. Money is the outcome. Judgment is the goal.

If an attacker can sway your thinking, even briefly, they can steer your actions. That single moment is all they need.

Give that careful thought.

No system was compromised, no password was cracked, and no firewall was bypassed.

A decision was made, and it changed everything.

How Modern Manipulation Works

Every successful manipulation, whether it involves a scam, deception, or engineered influence, follows a pattern. It may look different on the surface, but the underlying structure remains the same.

STAGE 1 — ATTENTION CAPTURE

The attacker needs to capture your attention quickly. **This is achieved by:**

- urgency ("Act now")
- fear ("Your account is compromised")
- attraction ("I've been looking for someone like you")
- authority ("This is the bank… law enforcement… your employer") They are not trying to inform you. They are trying to disrupt your usual way of thinking.

STAGE 2 — EMOTIONAL ACTIVATION

Once they capture your attention, they evoke emotion. Emotion is the doorway.

Fear makes you act quickly. Hope leads you to overlook risk. Love leads you to trust. Panic disables logic.

The goal isn't to persuade you with facts. It's to shift you from thinking to reacting.

STAGE 3 — TRUST FABRICATION

They are now building credibility. **They may:**

- use official language
- reference known institutions
- mimic someone you know
- build emotional connection
- present "evidence" that seems real

AI has transformed everything. Tasks that once required skill and effort can now be created instantly.

Messages are refined. Voices sound genuine. Responses are instant. Trust is built.

STAGE 4 — PRESSURE AND URGENCY

Now, the turning point arrives.

You are told: "This must be done now." "Don't tell anyone." "You are the only one who can fix this." "If you wait, it will be worse."

This isn't about speed; it's about solitude.

They are trying to take away your ability to verify, think, consult, and pause. If you pause, you break the sequence.

STAGE 5 — ACTION EXTRACTION

This is the last stage. You are instructed to:

- send money
- provide information
- transfer funds
- click a link
- comply with instructions
- make a decision you would not normally make

Here is the key: by the time you reach this stage, the decision seems justified. That's the danger.

REAL WORLD SCENARIO: THE CALL THAT CHANGES EVERYTHING

A man receives a call. The voice on the line sounds like his daughter, and she is crying.

Dad... I'm in trouble... Please don't hang up...

Another voice comes on. "This is law enforcement. Your daughter has been involved in an incident. If you want to help her, you need to act immediately."

The father's heart pounds. He isn't considering verification; he's focused on his child.

He is told to send the money immediately. At that moment, he has two options:

PATH 1 — REACTION

He sends the money. He realizes later that it wasn't real. The damage is already done.

PATH 2 — DEFENSIVE INTELLIGENCE

He pauses, hangs up, then calls his daughter directly to confirm.

The whole situation falls apart.

The Difference Between Damage And Control

The difference wasn't about intelligence or education.

The difference came down to one decision: pause… or react?

Why This Matters More Than Ever

We are entering a time when messages will look real, voices will sound real, conversations will feel real, and urgency will feel justified.

But reality and authenticity are no longer identical.

The Core Shift This Book Demands

This book isn't asking you to become paranoid. It's asking you to be disciplined.

Paranoia sees everything as a threat. Discipline checks first. Paranoia reacts emotionally. Discipline pauses deliberately. Paranoia isolates, yet discipline provides confirmation.

The First Rule Of Defensive Intelligence

Before responding to anything that creates pressure, urgency, fear, or emotional intensity, pause.

Not for very long. Just enough time to think.

That pause isn't a weakness. That pause is control.

Field Application: What You Do Starting Today

Starting now:

When you receive:

- an urgent message
- a financial request
- an emotional appeal

- a threatening message
- a "too good to be true" opportunity You don't respond immediately.

You use this approach:

The 10 Second Defensive Reset

- What am I feeling right now?
- What is being asked of me?
- Why is there urgency?
- Can this be verified independently?
- What happens if I wait?

If you can't answer those questions clearly, you do nothing.

Final Word For This Chapter

The biggest threat today isn't artificial intelligence. It's unprotected human-decision-making when influenced by artificial forces.

And the purpose of this system, this book, is simple: to ensure that when the moment comes… you are not caught off guard, you are not hurried into regret, and you are never outmaneuvered in your own mind.

CHAPTER 1 TAKEAWAY

The attack begins in your mind.

The defense begins with your pause.

CHAPTER 2

Synthetic Deception: When Reality Can Be Manufactured

There was a time when seeing was believing. If you saw a video, it was genuine. If you heard a voice, it was authentic. If someone spoke clearly, confidently, and intelligently, you believed they were who they claimed to be. That period has ended.

We are now living in an era where reality can be created, on demand, at scale, and with increasing accuracy. The boundary between what is real and what is generated has become thin, and in many cases, invisible.

This isn't just a theory or future speculation. This is happening now.

What "Synthetic Deception" Really Means

Synthetic deception uses AI and digital tools to create content like voices, images, videos, and conversations that appear real but are not.

It's not just fake. It's convincingly fake. And that difference is what makes it risky.

Because the human brain isn't wired to question realism; it's wired to accept it. When something appears real, sounds real, and feels real, we instinctively accept it as truth. That natural impulse is now being exploited against people.

The Four Pillars Of Synthetic Deception

To comprehend the threat, you must understand how it's constructed.

1. VOICE CLONING: WHEN ANYONE CAN SOUND LIKE ANYONE

A person's voice used to serve as a mark of identity. Now, it exposes their vulnerability.

With only seconds of audio, AI can mimic tone, cadence, and speech patterns.

That means:

- a parent can receive a call that sounds exactly like their child
- an employee might receive instructions that sound exactly like their boss.
- a victim can hear a voice they trust and react without hesitation. The real danger isn't the technology itself, but the emotional response it provokes.

When you hear a familiar voice in distress, your brain doesn't slow down. It speeds up.

And that acceleration is where errors occur.

2. DEEPFAKE VIDEO: SEEING WITHOUT KNOWING

Video used to be one of the most reliable forms of evidence. Now, it can be altered, generated, or manipulated to show things that never actually happened.

Faces can be mapped. Expressions can be recreated. Movements can be simulated.

Someone can seem to say something they never actually said. A situation can seem real even if it never existed.

And here's the main issue: most people aren't trained to recognize it.

They depend on instinct. And instinct alone isn't sufficient.

3. AI GENERATED COMMUNICATION: THE PERFECT MESSAGE

There was a time when scams were simple to identify: poor grammar, odd phrasing, and clear mistakes.

That moment has passed. **Today, messages can be:**

- grammatically perfect
- emotionally intelligent
- context aware
- personalized

AI can mirror tone, adapt responses, sustain long conversations, and build trust over days or weeks.

This is no longer just random messaging. This is a crafted conversation.

4. Synthetic Identities: People Who Don't Exist

Profiles can be made to look entirely real:

- unique photos
- consistent backstories
- authentic-looking social activity
- networks that strengthen credibility

These identities can establish relationships, earn trust, ask for help, and influence decisions.

Since they're not real people, they can't be held accountable.

The Authenticity Gap

Here is the key concept in this chapter:

What appears real, sounds real, and feels real is no longer proof that it actually is.

This is the gap in authenticity.

If you don't adapt to this reality, you'll keep judging situations based on outdated assumptions.

REAL WORLD SCENARIO: THE EXECUTIVE CALL

A finance officer receives a call. The voice is their CEO.

Calm. Direct. Familiar.

"I need you to process a transfer right away. This is time sensitive. I'll explain later."

The voice is right. The tone is right. The authority is right. Everything looks good, except for one thing. It's not authentic. The voice has been cloned. The request is fraudulent.

What Happens Next

PATH 1 — TRUST THE VOICE

- Transfer completed
- Funds lost
- Realization occurs after

PATH 2 — APPLY DEFENSIVE INTELLIGENCE

- Pause
- Verify via an alternative method
- Verify that the request is false.

The situation concludes before damage happens.

Why People Still Fall For This

Not because they lack intelligence, but because they depend on signals that used to be reliable.

Voice used to define identity. Video used to serve as evidence. Clear communication used to establish credibility.

Those assumptions are no longer reliable.

The Moment Of Decision

Every artificial deception result in a single point: the moment when you must decide whether to act.

That moment is when everything matters.

The Defensive Shift You Must Make

You need to shift from: "I recognize this, so it must be real." To: "I recognize this, but I will verify it anyway."

That shift is what separates control from exposure.

Field Application: How To Verify In The Real World

Verification should be conducted in multiple layers.

LAYER 1 — BREAK THE CHANNEL

Never confirm information using the same source it originated from.

If you receive:

- a call → hang up and then call back using a known number.
- a message → verify through a separate platform
- an email → verify via official contact points

LAYER 2 — REQUIRE FRICTION

Real situations can endure delays. Manipulation cannot.

If someone says, "Do this now." "Don't tell anyone." "There's no time to verify."

That is not an emergency. That is a warning.

LAYER 3 — ASK CONTROL QUESTIONS

Ask questions only the real person would know.

If responses are delayed, vague, or deflected, that is data.

LAYER 4 — TRUST STRUCTURE, NOT FEELING

Feelings can be manipulated. Structure cannot. If the situation:

- breaks protocol
- bypasses the normal process
- introduces a sense of unusual urgency

It needs to be verified. No exceptions.

REAL WORLD SCENARIO: THE FAMILY EMERGENCY CALL

A mother gets a call. It sounds like her son. He is crying. "Mom, please... I need help..."

Another voice cuts in: "He's in trouble. You need to send money immediately."

Her instinct is to act, but instead, she pauses.

She hangs up and calls her son directly. He answers, and he's fine.

The whole situation vanishes in seconds.

What Just Happened

The attack was based on:

- emotional shock

- voice familiarity
- urgency
- isolation

It was unsuccessful due to:

- one pause
- one verification

The Cost Of Not Adapting

If people keep acting as if:

- voices verify identity.
- video proves the truth.
- messages demonstrate authenticity.

They will keep being targeted.

Not because they are careless, but because they are relying on outdated rules in a new environment.

The Second Problem No One Talks About

The real risk isn't just that synthetic deception can fool you. The greater danger is this:

It teaches you to rely on the wrong signals.

Once your logic is compromised, there's no need for harsh attack. You can be guided gently.

The Rise Of "Guided Decisions"

Modern manipulation isn't about forcing action; it's about guiding you toward a decision that seems like your own.

The Guided Decision Model

1. Establish familiarity
2. Create comfort
3. Introducing a subtle need.
4. Apply a gentle sense of urgency
5. Allow the victim to justify the action

By the time the action occurs, the victim feels responsible.

REAL WORLD SCENARIO: THE LONG GAME ROMANCE + AI ASSISTED

A man connects with someone online. The conversation is steady, considerate, and emotionally intelligent.

Weeks become moments of connection.

Then one day: "I didn't want to ask… but I'm in a difficult situation…"

No pressure, no demands, just a story.

He offers help and believes the trust is genuine. But the whole interaction was planned.

CONVERGING DECEPTION

Multiple signals working in unison to remove doubt.

You don't evaluate the pieces individually. Instead, you step back and ask:

- Why does everything seem to be pointing in the same direction?
- Who stands to benefit if I act now?
- Has this been independently confirmed?

THE G3TI SYNTHETIC DETECTION FRAMEWORK

S.D.V.U. MODEL

S — **Source:** Where does this come from?

D — **Delivery:** How did it get to me?

V — **Verification:** Has it been confirmed outside this interaction?**U**

— **Urgency**: Why does this need to be done now?

RULE: If urgency escalates while verification drops… you are probably being manipulated.

Field Application: Realtime Decision Check

Ask yourself:

- If this were fake, how would it appear?
- What part of this can't be independently proven?

- What would change if I waited 10 minutes? Manipulation relies on timing. Truth can endure delays.

The Silent Weapon: Isolation

"Don't tell anyone." "This is confidential." "You need to handle this yourself."

This is not an instruction. This is control.

DEFENSIVE RULE: If you are told not to verify, that is the clearest sign that you need to verify.

The Cost Of Over Confidence

The most vulnerable person isn't the uninformed; it's the one who believes, "This would never happen to me."

Final Strategic Truth For This Chapter

You are not expected to distrust everything. You are being taught to trust properly.

Blind trust can be risky. Disciplined trust holds great strength.

The New Standard

From this point forward, Trust is no longer given based on appearance. It is established through verification.

Field Exercise: Start Today

1. Create a family verification plan.
    - code word

   - secondary contact method
2. Establish a rule of no urgent money requests.
   - no transfers without verification.
3. Practice channel breaking
   - always confirm outside the original message.
4. Learn to pause when under pressure
   - urgency signals you to slow down.

CHAPTER 2 TAKEAWAY

In the AI era, realism is no longer proof; verification is.

The question is no longer: "Does this look real?"

The question is: "Has this been verified?"

CHAPTER 3

Why Smart People Get Manipulated

Psychology, Ego, Emotion, and the Breakdown of Decision-Making

There is a harmful myth that has been spread for years: "Only gullible people get scammed."

That is false. It's not just incorrect, but dangerously misleading. Because it causes intelligent, educated, capable people to lower their guard, and that's exactly what makes them vulnerable.

The Truth No One Wants To Say

Smart people aren't immune to manipulation. In many cases, they make easier targets.

Why?

Because intelligent people:

- trust their judgment
- move quickly through decisions swiftly.

- believe they can "figure things out"
- don't want to seem unsure
- often function independently

These are strengths. But under pressure, those same strengths can become vulnerabilities.

The Real Vulnerability Is Not Intelligence

It is the moment your thoughts are disrupted.

Manipulation does not overpower intelligence. It sidesteps it.

The Decision Hijack

Every manipulation event includes a moment when your normal thinking process is disrupted. This is known as the Decision Hijack.

It happens when:

- emotion spikes
- urgency heightens
- information becomes unclear
- pressure is applied
- time feels compressed. At that moment, the brain shifts. **You shift from:**
- analysis → reaction
- logic → emotion
- verification → action

And once that shift occurs… the outcome is already underway.

The Five Human Pressure Points

Every person, regardless of intelligence, has these.

1. FEAR

Fear is the quickest override.

- "Your account has been compromised."
- "You are being investigated."
- "Your loved one is at risk." Fear doesn't question. It takes action.

2. URGENCY

Urgency removes time.

- "Do this now."
- "You don't have time to wait."
- "This will get worse if you delay."

Time is your most powerful defense. Urgency aims to strip it away.

3. AUTHORITY

People are conditioned to obey authority.

- law enforcement
- banks
- supervisors
- government agencies

When authority is established, resistance decreases.

4. TRUST

Trust is strong and can be exploited.

- relationships
- familiarity
- emotional connection

Once trust is built, verification typically fades away.

5. ISOLATION

Isolation is the hidden weapon.

- “Don’t tell anyone.”
- “Handle this yourself.”
- “This is confidential.”

When you're isolated, your thinking isn't challenged anymore.

REAL WORLD BREAKDOWN: THE SMART PROFESSIONAL

A well-educated professional gets an email. It seems to be from a vendor they frequently work with.

The tone is appropriate. The formatting appears neat. The request is sensible.

“Please update your payment information for future invoices.”

No urgency. No apparent red flags.

They comply.

Later, they realized that the email was fraudulent, the account had been altered, and the funds were redirected.

What Happened?

This wasn't carelessness; it was deliberate manipulation.

- familiarity was employed
- context was accurate.
- pressure was subtle.
- confidence replaces verification

The Lesson

Just because there are no red flags doesn't mean something is true.

The Role Of Ego

This is where many people struggle and won't admit it.

Ego has a role. Not loud arrogance, but internal confidence.

The belief that:

- "I would catch this."
- "I know what I'm doing."
- "I'm not easily fooled."

That belief reduces friction. And friction is what keeps you safe.

THE G3TI EGO CHECK

When you think, "This is obvious."

You need to ask: "What am I not checking because I think I already understand this?"

Emotional Narrowing

When emotion increases, you're thinking narrows. You lose sight of options. You concentrate on:

- the problem
- the pressure
- the quick fix

You stop asking:

- what else could this be?
- what am I missing?
- what happens if I wait?

REAL WORLD SCENARIO: THE FAMILY PANIC MOMENT

A grandmother receives a message: "Grandma, I'm in trouble. I need help."

It seems to be from her grandchild. **She feels:**

- fear
- urgency
- responsibility

Her focus sharpens solely on one goal: Help immediately. **She does not:**

- verify
- question
- delay

Because, emotionally, delay feels like harm.

The Breakdown

Emotion drove urgency. Urgency bypassed verification. Verification could have prevented the loss.

The "Helpful Person" Trap

Good people become targets. **People who:**

- want to help
- want to fix problems
- respond quickly
- care about others

These qualities are not flaws. But they are expected.

And anything predictable can be taken advantage of.

THE SHIFT

From: "I need to help right now."

To: "I need to help correctly."

Cognitive Overload

Another method used is overload. **You are provided with:**

- too much information

- conflicting details
- quick instructions

Your brain focuses on keeping up, not assessing.

WHEN THIS HAPPENS

You need to slow down, not speed up.

THE G3TI DECISION STABILITY FRAMEWORK

S.T.A.B.L.E. MODEL

S — **Slow down the moment:** Do not match their urgency.

T — **Think clearly:** What is actually being asked?

A — **Analyze pressure:** Why does this feel urgent?

B — **Break the channel:** Verify outside the current interaction.

L — **Look for patterns:** Does this follow known manipulation tactics?

E — **Execute deliberately:** Choose action, not reaction.

REALWORLD SCENARIO: THE CONFRONTATION TRAP

A person receives a message accusing them of wrongdoing. The tone is aggressive, and the language is provoking.

The goal is straightforward: provoke a reaction.

PATH 1 — REACTION

- respond emotionally

- escalate conflict
- create consequences

PATH 2 — DEFENSIVE INTELLIGENCE

- pause
- don't engage right away.
- assess intent
- respond strategically, or not at all.

The Lesson

Not every message warrants a reply. Every reply has an impact.

The Regret Window

There is always a window between:

- receiving the message
- taking action

That window may be seconds long. But it exists. This Book Helps You Expand That Window.

Because inside that window... lies your power.

Final Truth Of This Chapter

You are not being outsmarted. You are being engaged at this very moment **where:**

- your emotions are running high.
- your time feels compressed.
- your thinking is limited.

And at that moment... anyone can be vulnerable.

Final Close

The real question isn't if you're smart enough to avoid manipulation, but if you're disciplined enough to pause when it matters most.

Because when emotion rises and pressure is applied, intelligence doesn't protect you—discipline does.

CHAPTER 3 TAKEAWAY

Smart people are not immune to manipulation. They are vulnerable during moments of pressure, just like everyone else.

PART II

THE HUMAN DECISION UNDER ATTACK

CHAPTER 4

The Manipulation Playbook

How They Think, How They Move, and How They Win, Until You See It

Most people think manipulation happens randomly. It does not.

It is structured. It is repeatable. And most importantly, it is trained.

The First Truth You Must Accept

The person on the other end isn't guessing. They aren't just "trying something."

They are following a playbook.

A playbook refined through countless interactions, failures, successes, and adaptations.

That means when you are being targeted... You're not just in a conversation; you're in a process.

THE G3TI MANIPULATION SEQUENCE

Every effective manipulation follows this sequence:

1. **Select the Target**

2. **Study the Target**
3. **Initiate Contact**
4. **Build Control**
5. **Apply Pressure**
6. **Extract Action**
7. **Disengage or Repeat**

If you understand this sequence, you can interrupt it.

STAGE 1 — TARGET SELECTION

You weren't chosen at random. You were simply available. **Targets are chosen based on:**

- accessibility (website, phone, email)
- behavior (responsive, active, engaged)
- role (financial authority, caregiver, leader)
- emotional profile (lonely, helpful, stressed, busy)

Reality Check

You don’t need to appear vulnerable. You just need to be accessible.

Field Insight

Attackers search for:

- people who reply promptly
- people who answer unknown calls
- people who share personal details publicly

- people who seem trustworthy or responsible

STAGE 2— TARGET STUDY

Information is collected before contact. Not always thoroughly, but sufficiently.

This can include:

- social media
- job role
- family ties
- communication style
- routines

This Matters

When they contact you, they don't sound random; they sound informed.

And that creates trust.

REAL WORLD SCENARIO: SOCIAL MEDIA INTELLIGENCE

Someone posts: "Heading out of town for the week!"

Days later, they get a message: "Hi, this is your neighbor. There's an issue with your property..."

The message mentions timing, location, and context. It feels genuine. Because it was constructed with real information.

STAGE 3 — INITIATING CONTACT

This is the starting point of the interaction. The main goal is straightforward: to get a response.

Common Entry Points

- text message
- phone call
- email
- social media message
- dating platform
- job platform

Tactics Used

- curiosity ("Are you available?")
- concern ("We noticed suspicious activity")
- opportunity ("You've been selected")
- familiarity ("It's me")

Rule

The initial message isn't the attack; it's the entrance.

STAGE 4 — BUILDING CONTROL

This is where most people lose because nothing seems wrong.

What Happens Here

- conversation continues

- tone becomes relaxed
- trust starts to develop
- resistance decreases

This Is Where Ai Is Most Powerful

Because:

- responses are instant
- tone remains consistent.
- language is polished.
- emotional cues are reflected.

The conversation feels natural because it is designed that way.

REAL WORLD SCENARIO: THE TRUST BUILD

A person receives a message on a professional platform: "Hi, I came across your profile. I'd like to connect."

The conversation unfolds over several days.

Topics include:

- work
- goals
- shared interests

Nothing inappropriate. Nothing urgent.

Then gradually say: "I have an opportunity I think you'd be perfect for…"

The change has started.

STAGE 5 — PRESSURE APPLICATION

Now the tone shifts. Subtly at first, then more clearly.

Forms Of Pressure

- urgency ("This must be done now")
- exclusivity ("Only you can do this")
- fear ("There will be consequences")
- emotional appeal ("I need your help")

The Key Strategy

Pressure is applied only after trust is established.

REAL WORLD SCENARIO, FINANCIAL REQUEST

After weeks of communication: "I didn't want to ask, but I'm in a difficult situation..."

The tone is uncertain, not forceful. This reduces resistance. Then comes the moment: "I really need your help right now."

This Is The Moment

If you act without verifying... The sequence is complete.

STAGE 6 — ACTION EXTRACTION

This is where the outcome is guaranteed.

What They Want

- money

- access
- information
- compliance
- reaction

Important

By this point, the request feels:

- justified
- necessary
- time sensitive

That feeling is intentional; it stems from everything that happened before.

STAGE 7 — DISENGAGE OR REPEAT

Once the action has been completed:

- they vanish
- they escalate
- they go through the cycle again.

Repeat Strategy

If successful, they might:

- request more
- increase urgency
- strengthen emotional bond

Why People Get Caught Again

Because they are now invested emotionally and financially.

The Multilayer Attack

This is where things get more sophisticated.

Example

You receive:

- an email
- a follow-up call
- a confirming message

Each piece complements the other.

THIS CREATES A FALSE REALITY

Because multiple signals establish belief.

This Is Called

Layered Confirmation

Defensive Response

You need to step outside the whole system, not just evaluate each piece individually.

THE G3TI PLAYBOOK INTERRUPTION MODEL

I.N.T.E.R.R.U.P.T.

I — Identify the stage: Where am I in the sequence?

N — Notice pressure: What is being applied?

T — Take a pause: Do not act immediately.

E — Exit the channel: Verify externally.

R — Reframe the situation: What if this is manipulation?

R — Reach out: Consult a second source.

U — Understand the outcome: What happens if I act?

P — Protect your position: Do not expose information.

T — Take controlled action: Only after verification.

Real World Scenario: The Bank Call" Playbook

A person gets a call: "This is your bank. We detected suspicious activity."

Playbook In Action:

- attention captured
- fear introduced
- authority established
- urgency has been applied.

Request

"Transfer your funds to protect your account."

Interruption

- hang up
- call the bank directly.
- verify

Result

Attack fails.

The Toughest Truth In This Chapter

You aren't being targeted because you're weak.

You're being targeted because:

- you are reachable
- you are responsive
- you are human

And humans can be affected.

Final Strategic Shift

You need to stop asking: "Is this real?"

Then start asking: "What stage of the playbook is this?" Once you recognize the pattern, the pressure diminishes.

Final Close

Every manipulation proceeds in a sequence.

Once you observe the sequence, you are no longer reacting within it. You are managing your position outside of it.

CHAPTER 4 TAKEAWAY

You are not in a random situation. You are following a structured playbook. Once you recognize it, you can break free.

CHAPTER 5

The Emotional Override

How Fear, Love, Urgency, Shame, and Anger Take Control—And How You Take It Back

There is a moment, small, almost invisible, where everything changes. You receive a message. A call. A situation.

And suddenly your heart rate increases. Your thoughts speed up. Your focus narrows. You feel something strongly.

And in that moment, you are no longer thinking clearly.

This Is the Emotional Override

It is the point where emotion becomes stronger than logic. Not because you are weak. Because you are human.

The Truth Most People Don't Understand

Manipulation does not begin with a request. It begins with a feeling. Because if someone can control what you feel, they can influence what you do next.

What Happens Inside the Mind

When emotion spikes, the brain shifts priority. It moves from:

- analysis → survival
- reasoning → reaction
- patience → immediacy

You are no longer asking, "What is true?"

You are asking, "What do I need to do right now?" That shift is where manipulation lives.

The Five Emotional Overrides

Every manipulation strategy relies on at least one of these. Many employ multiple simultaneously.

1. FEAR — THE FASTEST OVERRIDE

Fear strikes instantly. It doesn't negotiate or hesitate. It simply acts.

What Fear Sounds Like:

- "Your account has been compromised."
- "You are being investigated."
- "Your loved one is in danger."
- "If you don't take action now, this will only get worse."

REAL WORLD SCENARIO: THE ACCOUNT LOCKDOWN

A man gets a message: "Suspicious activity detected. Your account will be frozen within minutes unless you act."

His thinking narrows. He is no longer analyzing; he's reacting to avoid loss.

What He Misses:

- the source is unverified
- the sense of urgency is artificial.
- the instruction is outside the normal process.

The Lesson

Fear is not knowledge. It is pressure.

2. URGENCY — THE TIME COLLAPSE

Urgency hinders your ability to think because thinking needs time.

What Urgency Sounds Like:

- "Right now."
- "Immediately."
- "Don't hesitate."
- "You only have a few minutes."

REAL WORLD SCENARIO: THE TRANSFER REQUEST

An employee receives instructions: "This must be completed before the end of the hour. No delays."

The request seems legitimate, but urgency removes the pause.

The Truth

Real processes enable verification. False ones prioritize speed.

3. LOVE AND CONNECTION: THE SOFT OVERRIDE

This is one of the most dangerous ones because it doesn't feel like pressure; it feels like trust.

What It Sounds Like:

- "I'm here for you."
- "I trust you."
- "You're the only person I can count on."
- "I didn't want to ask anyone else."

REAL WORLD SCENARIO: THE LONG CONNECTION

Weeks of conversations. Shared stories. Consistent communication. Emotional investment.

Then: "I need help."

The Shift

The decision is no longer logical; it is relational.

The Lesson

Connection can be genuine, but it can also be built.

Shame — The Silent Override

Shame is strong because it silences people. What It Sounds Like

- "You are responsible for this."
- "You need to fix this."

- "Don't tell anyone."
- "Handle this yourself."

REAL WORLD SCENARIO: THE MISTAKE TRAP

Someone is told: "There has been a breach connected to your account. This could become serious."

They feel responsible, embarrassed, and afraid of exposure.

What Happens

They try to fix it on their own. Without verification.

The Lesson

Shame causes isolation. Isolation removes safety.

4. ANGER — THE REACTION TRAP

Anger is different. It doesn't make you obey; it makes you react.

What It Sounds Like

- "accusations"
- "insults"
- "provocation"
- "disrespect"

REAL WORLD SCENARIO: THE PROVOCATION MESSAGE

"You believe you're smarter than everyone. Let's see." The goal isn't just information; it's a reaction.

What Happens

- "you respond emotionally"
- "you escalate"
- "you expose yourself "

The Lesson

Not every attack is designed to make money. Some are intended just to provoke a reaction.

The Combination Attack

The most powerful manipulations involve emotions.

Example

Fear + urgency + authority: "This is law enforcement. There is a situation involving your family. You need to act now."

Why This Works

Each emotion strengthens the other.

- fear creates urgency
- urgency hinders critical thinking.
- authority reduces resistance

The G3TI Emotional Override Interrupt

This is one of the most crucial tools in this entire system.

E.M.O.T.I.O.N. Model

E — Evaluate the feeling: What am I feeling right now?

M — Measure intensity: Is this stronger than normal?

O — Observe the trigger: What caused this reaction?

T — Take a pause: Do not act yet.

I — Identify the ask: What am I being asked to do?

O — Offset the pressure: Slow down the situation.

N — Navigate deliberately: Choose your next move with purpose.

The 3-Second Check That Can Save You

When emotions rise, ask:

- Why do I feel this way so intensely right now?
- Who gains if I act now?
- What occurs if I wait?

If you can't answer clearly, then do nothing.

REAL WORLD SCENARIO: THE BREAKPOINT

A woman gets a message: "I need help. Please don't ignore this."

Her emotions intensify. She feels urgency, concern, and responsibility.

Old Response

- respond immediately
- connect emotionally
- move toward taking action

New Response

- pause
- assess
- verify

Result

Control remains in place.

The Regret Line

Every emotional override results in a line. Once you cross that line, the action can't be undone.

Money transferred. Information shared. Words spoken. Decisions made.

The Strongest Truth in This Chapter

Emotion itself is not the enemy. It's uncontrolled emotion that is.

Final Strategic Shift

You don't eliminate emotion. You manage it. You feel it, but you don't let it decide.

Final Close

When emotion flares up, it's not the time to act. Instead, it's the time to pause.

Because when emotion takes over, you lose control. And when you're not in control, someone else is.

CHAPTER 5 TAKEAWAY

If they can manipulate your emotions, they can influence your actions.

Manage your emotions, and you direct the outcome.

CHAPTER 6

The Regret Point

The Exact Moment Before an Irreversible Action, and How to Stop It Every Time

There is a moment, before every mistake, every loss, and every decision someone wishes they could take back. It is small. It is fast. And most people don't recognize it until it's gone.

That moment is called the Regret Point

What Is the Regret Point?

It is the final moment before you take an irreversible action.

- before you send the money
- before you click the link
- before you share the information
- before you respond emotionally
- before you confront, accuse, or escalate

It is the line between control and its consequences.

The Truth About Regret

Regret does not stem from the situation. It stems from the decision made under pressure.

People don't say, "I regret what happened."

They say, "I knew something felt off… but I did it anyway."

Why People Cross the Regret Point

Not because they are careless. Because they are:

- overwhelmed
- rushed
- emotional
- isolated
- trying to fix something quickly

At that moment, they choose relief over clarity.

The Regret Pattern

Every regret follows this pattern:

- Something triggers emotion
- Pressure is applied
- Time is compressed
- Thinking narrows
- Action is taken
- Realization follows

The Most Important Part

There is always a gap between steps 4 and 5. That gap is your last opportunity.

REAL WORLD SCENARIO: THE TRANSFER

A business owner receives a message: "We need to finalize this transaction today.

Please send the updated payment right away."

Everything appears normal. The language is professional, and

the request is familiar.

But something feels slightly off. This Is the Regret Point

The moment the thought appears: "This feels different…"

Path 1 — Ignore the Signal

- proceeds with transfer
- realizes later it was fraudulent
- loss occurs

Path 2 — Honor the Signal

- pauses
- verifies independently
- confirms the request is false

The Difference

Not intelligence. No experience. One decision: Pause… or proceed.

The "Just This Once" Trap

This is one of the most dangerous thoughts: "This is probably fine… just this once."

That thought bypasses your usual standards.

What It Does

- lowers your guard
- removes verification
- accelerates action

The Truth

The moment you say "just this once," you are no longer acting with discipline. You are acting with emotion.

The G3TI Regret Interrupt Protocol

This is one of the most important tools in this book.

R.E.G.R.E.T. Model

R — **Recognize the moment:** Something feels off. Acknowledge it.

E — **Exit immediacy:** Do not act right now.

G — **Get distance:** Step away mentally or physically.

R — **Reassess facts:** What do I actually know?

E — **Engage verification:** Confirm through independent channels.

T — **Take controlled action:** Only after clarity is established.

The 10 Minute Rule If the action is:

- financial
- emotional
- irreversible

You wait. Minimum: ten minutes.

Why This Works

Because manipulation depends on speed, time restores thinking.

REAL WORLD SCENARIO: THE EMOTIONAL MESSAGE

A person receives: "I can't believe you did this. You need to respond right now."

The message is aggressive and provoking.

Regret Point

The urge to respond immediately.

Old Response

- reply emotionally
- escalate conflict
- create consequences

New Response

- pause
- do not respond immediately

- revisit with clarity

Result

No escalation. No regret.

The Body Signals You Must Pay Attention To

Your body will tell you before your mind catches up.

Warning Signs

- increased heart rate
- tension
- urgency
- mental pressure
- feeling rushed

These are not random. They are indicators that you are approaching the regret point.

The "Future Self" Check

Before acting, ask: "What will I think of this decision tomorrow?"

If you cannot answer with confidence, you are not ready to act.

The Cost of Crossing the Regret Point

Once crossed:

- money may be gone
- information exposed
- relationships damaged

- legal consequences created
- emotional distress begins And the Hardest Part

You often knew. You felt it. Yet you moved anyway.

The Power of Stopping

Stopping is not a weakness. It is control. Reality

Most people believe strength means acting quickly. True strength means holding your ground under pressure.

The G3TI Decision Lock

This is your final safeguard.

Rule

No irreversible action without:

- pause
- verification
- clarity No exceptions.

REAL WORLD SCENARIO: THE FINAL SECOND

A person is about to send money. They are seconds away, and everything is in place.

Then they stop, not because of fear but because of discipline. They verify. The situation collapses.

This Is What This Book Is Building

Not awareness. Interruption.

Final Strategic Truth

You do not need to predict every threat or understand every tactic. You only need to control one moment: the moment before you act.

Final Close

Every mistake has a moment before it happens. Every loss has a second before it becomes real.

And every person has one opportunity to stop.

Once you act under pressure, you are no longer deciding. You are reacting. And reaction is where regret begins.

CHAPTER 6 TAKEAWAY

The most powerful move you have is not action; it is the ability to pause before acting.

PART III

VERIFICATION, COUNTER MOVES, AND DEFENSIVE SKILL

CHAPTER 7

Verification Is The New Self-Defense

How to Prove Truth in a World Where Anything Can Be Faked

There was a time when protection meant staying physically aware. Lock

your doors. Watch your surroundings. Be mindful of who is nearby.

Today, protection has changed. You can be sitting in your own home, on your own phone, talking to someone who sounds like someone you trust... and still be under attack.

The New Form of Self-Defense

Self-defense is no longer just physical. It is cognitive, digital, and behavioral.

At the center of it is one word: Verification

The Shift That Must Happen

In the past, people asked, "Does this look real?" That question is no longer adequate.

The question now must be: "Has this been verified?"

Why Verification Is Now Non-negotiable

Because:

- voices can be cloned
- videos can be generated
- messages can be perfected
- identities can be fabricated
- conversations can be engineered **This means:**
- Appearance is no longer proof.

The Most Dangerous Assumption

"This looks legitimate."

That assumption has cost people dearly:

- money
- careers
- relationships
- reputations
- safety

Legitimacy is no longer based on how something looks. It is based on whether it can be independently verified.

The Verification Gap

Most people do not verify. Not because they don't want to, but because they don't know how or don't think they need to.

What People Do Instead

- trust the message

- trust the voice
- trust the format
- trust the urgency

What They Should Do

Break the interaction. Step out of the situation. Confirm independently.

The G3TI Verification Framework

This becomes one of the most powerful systems in the entire book.

V.E.R.I.F.Y. Model

V — Validate the source: Who is this really from?

E — Exit the channel: Do not verify within the same interaction.

R — Reconfirm identity: Use known, trusted contact points.

I — Inspect the request: Does this follow normal behavior or process?

F — Find independent proof: Can this be confirmed outside this message?

Y — Yield only when certain: No action without clarity.

The First Rule of Verification

You never verify a situation in the same environment where it was created.

Why This Matters

If the message is fake, the confirmation it contains will also be fake.

REAL WORLD SCENARIO: THE CONFIRMATION TRAP

A person receives an email: "Please confirm your account details to avoid suspension"

They click the link, see a professional-looking page, and enter their information.

What They Think

"I verified it, and it looked real."

What Actually Happened

They verified within a controlled environment created by the attacker.

The Lesson

Verification must occur outside the system.

The Three Layers of Verification

To operate at a high level, you must verify in layers—not just once.

Layer 1 — Source Verification

Who is contacting you?

- Is this a known number?
- Is this a verified email domain?

- Is this a trusted identity?

Layer 2 — Channel Verification

How did they contact you?

- Is this the normal method?
- Is this consistent with past communication?

Layer 3 — Context Verification

Does this situation make sense?

- Is this request normal?
- Is the timing logical?
- Is the urgency justified?

REAL WORLD SCENARIO: THE EXECUTIVE REQUEST

An employee receives the message: "Send me the updated files immediately. I'm in a meeting."

The message appears to originate from leadership.

Verification Process

- calls the executive directly
- confirms the request was never made

Result

Threat neutralized.

The Power of Breaking the Channel

This is one of the most important skills you will develop.

What It Means

You do not stay within the conversation. You step outside it.

Examples

- call instead of text
- use official contact information
- speak to someone directly
- verify through known channels

Why This Works

Because attackers control the original channel, they cannot control what happens outside of it.

The Friction Rule

Real situations can handle friction. Fake ones cannot.

If Someone Resists Verification

- "There's no time."
- "Just do it now."
- "Trust me."

That is not a request. That is a signal.

Rule

If verification is discouraged, the situation must be verified without delay.

Advanced Verification: Asking the Right Questions

Not all verification is equal. You must ask targeted questions.

Weak Questions

- "Is this real?"
- "Are you who you say you are?" These can be answered easily.

Strong Questions

- details only the real person would know
- context specific information
- verification tied to past interactions

Example

Instead of: "Is this you?"

Ask: "What did we discuss in our last meeting regarding this?"

The Speed vs. Truth Principle

Truth is stable. It does not collapse over time. Manipulation depends on speed.

Rule

The faster you are pushed to act, the slower you must move.

REAL WORLD SCENARIO: THE BANK CALL

"This is your bank. We need you to move your funds immediately."

Verification Response

- hang up
- call the bank using the official number
- confirm status

Result No loss.

The Cost of Skipping Verification

When verification is skipped:

- decisions are made blindly.
- actions are taken under pressure.
- consequences become irreversible

And the Hardest Part

It often feels right in the moment.

The Verification Habit

This is not a one-time action. It is a mindset. **You Train Yourself To:**

- pause before responding
- verify before acting
- question before trusting
- confirm before committing

REAL WORLD SCENARIO: THE FINAL CHECK

A person is about to send payment, and everything appears legitimate. They apply one step: they verify externally.

Result

The request is false. **The Difference** One step.

The Strongest Truth in This Chapter

Verification is not about distrust. It is about safeguarding.

Final Strategic Shift

You do not trust faster. You verify first. Then you trust.

Final Close

In a world where anything can be made to look real, your safety is not in what you recognize. It is in what you confirm. Verification is no longer optional. It is your defense.

CHAPTER 7 TAKEAWAY

Trust what you can prove, not what you are shown.

CHAPTER 8

The Human Firewall

Becoming the Barrier That Deception Cannot Penetrate

Systems fail. Software gets bypassed. Security tools get updated, yet still get defeated. Algorithms get outpaced.

But there is one line of defense that, when properly trained, cannot be easily exploited: a disciplined human mind.

What a Firewall Does

A firewall does not chase threats. It does not panic or react emotionally. It does one thing: It evaluates what is trying to pass through and decides what gets blocked.

The Human Firewall

A human firewall is a person who:

- does not react under pressure
- does not trust without verification
- does not get pulled into emotional manipulation
- does not move faster than their thinking
- does not surrender control of their decision-making

This Is Not About Intelligence

It is about discipline because intelligence can be bypassed. Discipline cannot.

The Shift From Target to Defender

Most people move through life passively. They receive messages, respond, and react.

A human firewall operates differently. It:

observe first

- analyze second
- respond last

The Human Firewall Mindset

You must adopt this principle: "Nothing gets through me without being evaluated."

What This Looks Like in Real Life

You receive a message. Instead of reacting, you ask,

- What is this trying to make me feel like?
- What is this trying to get me to do?
- What happens if I do nothing now?

The Five Attributes of a Human Firewall

1. Controlled Response

You do not match urgency. You control timing.

Most manipulation depends on you reacting quickly. When you slow down, you disrupt the system.

2. Emotional Discipline

You feel, but you do not act on feeling alone. You understand: Emotion is data, not a direction.

3. Pattern Recognition

You recognize:

- urgency patterns
- pressure language
- authority positioning
- emotional triggers

Once you see patterns, you are no longer part of the situation. You are observing it.

4. Verification Habit

You do not assume. You confirm every time.

5. Positional Control

You never give up control of your decision-making. This means:

a. you don't act under pressure

b. you don't follow unfamiliar instructions blindly

c. you don't allow others to dictate your timeline

REAL WORLD SCENARIO: THE HUMAN FIREWALL IN ACTION

An employee receives a message: "I need this done immediately. Don't

delay."

Normal Response

d. react
e. comply
f. move quickly

Human Firewall Response

g. pause
h. evaluate tone and request
i. verify through another channel

False request identified. Threat mitigated.

The "Energy Shift"

When you become a human firewall, your energy shifts.

You no longer:

- chase explanations
- justify your hesitation
- feel pressured to respond immediately

You operate from control, not from reaction.

The Silent Power of Non-Response

One of your strongest tools is this: not responding immediately. Manipulation depends on engagement. Without engagement, the system cannot progress.

Real World Scenario, The Bait Message

"You need to respond immediately, or this will escalate."

Human Firewall Response

No immediate reply.

Result

The pressure loses its effect.

The G3TI Human Firewall Framework

This becomes a signature model.

F.I.R.E.W.A.L.L. Model

F — **Filter input:** Not everything deserves attention.

I — **Identify intent:** What is being attempted?

R — **Regulate emotion:** Stay in control.

E — **Evaluate request:** Does this make sense?

W — **Withdraw from pressure:** Do not act immediately.

A — **Authenticate source:** Verify independently.

L — **Lock decision:** No action without clarity.

L — **Lead response:** You decide how and when to respond.

The Human Firewall Test

Before responding, ask:

- Am I reacting or responding?
- Do I feel pressured?
- Have I verified this?
- What happens if I wait?

If you cannot answer clearly, do nothing.

REAL WORLD SCENARIO: THE FAMILY MESSAGE

A message appears to be from a relative: "I need help right now."

Human Firewall Response

- pause
- call directly
- confirm

Result

No action was taken on false information.

The Difference Between Defense and Control

Defense reacts to threats. Control prevents them from succeeding.

The human firewall operates under control.

The Cost of Not Being One

Without this mindset:

- You react too quickly.

- You trust too easily.
- You move under pressure.
- You skip verification.

That is all an attacker needs.

The Strongest Truth in This Chapter

You do not need to understand every threat. You need to become the kind of person that threats cannot move.

Final Shift

From: "What is happening to me?"

To: "What is being attempted, and how can I control my response?"

Final Close

The strongest defense is not technology. It is a person who cannot be rushed, pressured, or moved without clarity.

When you become that person, you are no longer a target. You are a barrier.

CHAPTER 8 TAKEAWAY

Become the filter, not the receiver. Evaluate everything. React to nothing.

CHAPTER 9

Technical Wit For Everyday People

The Minimum Technical Edge You Need to Protect Yourself in a High-Threat Environment

You do not need to be a cybersecurity expert to protect yourself. You do not need to write code, understand complex systems, or become technical in the traditional sense.

But you do need something most people lack: technical wit.

What Is Technical Wit?

Technical wit is the ability to:

- recognize when something is off in a digital interaction
- understand how common attacks actually work
- apply simple, effective countermeasures in real time
- avoid being overwhelmed by technology while using it intelligently

This Is the Truth

Most attacks do not succeed because systems are weak.

They succeed because of people:

- don't understand what they're looking at
- trust what appears professional
- don't know what to check
- move too fast

This Chapter Fixes That

The First Principle: Digital Does Not Mean Trusted

Just because something appears on your phone, email, or computer does not make it legitimate.

What People Assume

- "It came through my email, so it must be genuine."
- "It looks official, so it must be genuine."
- "It has a logo, so it must be authentic."

What You Must Understand

All of that can be replicated.

The Second Principle: Small Details Reveal Big Truths

Most deception fails in the details, but only if you know where to look.

The G3TI Digital Inspection System

This becomes one of your most practical tools.

L.O.O.K. Model

L — **Link:** Where does this actually go?

O — **Origin:** Who really sent this?

O — **Oddities:** What feels slightly off?

K — **Known behavior:** Does this match how this person or company normally operates?

REAL WORLD SCENARIO: THE EMAIL

You receive the message: "Your account needs immediate attention."

Technical Wit Response

You do not click immediately. You inspect first.

Step 1 — Hover Over the Link

The visible text says one thing, but the actual link says another.

Step 2 — Check the Email Address

It looks official… but has subtle differences.

Step 3 — Check the Tone

It creates urgency.

Result

You do not engage.

The Password Problem

This is one of the most common weaknesses.

What People Do

- reuse passwords
- use simple combinations
- store them insecurely

What You Must Do

Use:

- long passphrases
- unique passwords per account
- password managers

Why This Matters

If one account is compromised, all linked accounts are at risk.

Multifactor Authentication (MFA)

This is no longer optional.

What It Does

It adds a second layer:

- something you know (password)
- something you have (device/code)

Why It Works

Even if your password is stolen, access remains blocked.

REAL WORLD SCENARIO, THE ACCOUNT TAKEOVER

A person receives a notification: "Your password has been changed."

Without MFA: Account compromised. With MFA, the attack was stopped.

The Link Trap

Links are among the most common attack methods.

Rule

You do not click links from:

- unknown sources
- unexpected messages
- urgent requests

Alternative

Go directly to the official website.

The Spoofing Reality

Attackers can make messages appear to come from:

- your bank
- your employer
- your contacts

This is called spoofing. It is effective because it exploits trust.

Defense

You do not trust appearances. You verify the source.

REAL WORLD SCENARIO, THE TEXT MESSAGE

"Your package is delayed. Click here to reschedule."

What People Do

Click.

What You Do

Check tracking directly on the official site.

Social Media, The Data Goldmine

Every post reveals information.

What Attackers Look For

- location
- routines
- relationships
- personal details

Why This Matters

This information is used to:

- build trust
- create believable stories
- target you more effectively

The Privacy Shift

You don't have to stop using social media. You have to use it strategically instead.

Rule

Do not share information that could be used against you in real time.

Device Security

You must:

- update your devices
- use lock screens
- avoid unknown downloads
- use secure networks

Why

Many attacks exploit outdated systems.

The Public Wi-Fi Trap

Public networks are not secure.

Rule

Do not access:

- banking
- sensitive accounts
- personal data

on unsecured networks.

The "Too Easy" Rule

If something feels:

- too easy
- too fast
- too convenient It must be examined.

REAL WORLD SCENARIO, THE JOB OFFER

"You've been selected. Onboarding is immediate. Just send your details."

Technical Wit Response

- verify company
- confirm recruiter
- check domain
- do not send sensitive info

The G3TI Digital Defense Stack

This becomes a system that people can remember.

D.E.F.E.N.D. Model

D — **Don't** click immediately: Pause.

E — **Examine** details: Look closely.

F — **Find** the source: Who sent this?

E — **Exit** the message: Verify externally.

N — **Never** rush: Time is your ally.

D — **Decide** with clarity: Then act.

The Biggest Mistake People Make

They think, "I'll be careful next time."

This is not enough. You need systems, habits, and automatic responses.

The Goal of This Chapter

Not to make you technical. To make you:

Technically aware, digitally disciplined, and practically dangerous to deception.

Final Truth

Technology is not the threat. Misuse of technology is.

Final Close

You do not need to master technology to protect yourself. You need to understand just enough to avoid being controlled by it.

Because in this environment, awareness is protection. And discipline is power.

CHAPTER 9 TAKEAWAY

You don't need to know everything; you just need to know what to check before you act.

CHAPTER 10

Evidence Before Emotion

How to Preserve Truth, Protect Yourself, and Respond Without Making Things Worse.

There are two ways people respond when something goes wrong.

They either:

- react emotionally
- respond strategically

Most people:

- panic
- delete messages
- confront immediately
- try to fix everything at once
- say things they can't take back

And in doing so, they destroy the very evidence that could have protected them.

The First Rule of This Chapter

When something feels wrong, you do not act first. You preserve first.

Why Evidence Matters

Evidence does three things:

- protects your position
- reveals the truth
- allows proper action later

Without evidence:

- your claims become harder to prove
- your options become limited
- your control decreases

What People Do Wrong and Why It Costs Them

MISTAKE 1 — DELETING TOO FAST

"I don't want to deal with this."

So they:

- delete the message
- block the contact
- erase the conversation

What They Just Did

They removed:

- timestamps

- language patterns
- contact data
- proof of interaction

Reality

You cannot prove what you cannot demonstrate.

MISTAKE 2 — IMMEDIATE CONFRONTATION

"Who is this? What are you doing?" What This Does

- alerts the attacker
- changes their behavior
- causes them to disappear
- escalates the situation

You Lose

- visibility
- control
- information

MISTAKE 3 — EMOTIONAL RESPONSE

- anger
- panic
- fear
- urgency

What Happens

You:

- say too much
- reveal information
- act too quickly

Result

You give away your leverage.

The Shift You Must Make

From: "I need to respond right now."

To: "I need to secure this situation first."

The G3TI Evidence First Protocol

This becomes one of the most powerful tools in the entire system.

E.V.I.D.E.N.C.E. Model

E — **Establish awareness:** Something is not right.

V — **Verify nothing yet:** Do not engage or confirm.

I — **Isolate the interaction:** Stop active communication.

D — **Document everything:** Screenshots, recordings, timestamps.

E — **Extract key details:** Who, what, when, how.

N — **Notify appropriately:** Only after securing information.

C — **Control your response:** No emotional reaction.

E — **Execute next steps:** With clarity and structure.

What You Must Document: No Exceptions

When something suspicious happens, capture:

- full message threads
- phone numbers
- email addresses
- usernames
- timestamps
- links (do not click — copy if possible)
- screenshots of profiles
- any attachments

Why This Matters

Because once it's gone, it's gone.

REAL WORLD SCENARIO: THE MESSAGE THAT DISAPPEARS

A person receives a suspicious message. They feel uncomfortable and delete it.

Later, they realize it was part of a larger problem.

Problem

There is no record, no proof, and no reference point.

Lesson

Never delete first.

The Power of Silence

You do not need to respond to everything.

In many situations, your strongest move is no immediate response.

Why

Because:

- it prevents escalation
- it preserves your position
- it allows you to think

REAL WORLD SCENARIO, THE PROVOCATION

"You'd better respond right now."

Old Response Immediate reply. **New Response**

Silence. Documentation. Control.

Result

You remain in position.

The Difference Between Response and Re-action

Reaction

- emotional
- fast
- uncontrolled

- short term

Response

- deliberate
- informed
- controlled
- strategic

The Information You Should Never Give Away

When under pressure, do not share:

- personal information
- account details
- financial data
- verification codes
- emotional state

Why

Information is leverage.

REAL WORLD SCENARIO: THE "HELP ME" MESSAGE

"I need your help. Could you confirm your information so I can process this?"

What They Want

- data
- confirmation
- access

What You Do

Nothing. Until verified.

The Reporting Decision

After documentation, you decide:

- is this a scam?
- is this harassment?
- is this fraud?
- is this a threat? **Then you report:**
- financial institution
- employer
- appropriate authority
- official reporting channels

But only after you secure your position.

The "Second Damage" Problem

Most people don't lose in the first event. They lose because of what they do afterward.

Second Damage Includes

- sending more money

- trying to fix things emotionally
- confronting without evidence
- exposing more information This chapter stops that.

The G3TI Control Triad

This becomes a central concept.

- **Preserve** — capture everything
- **Position** — do not engage emotionally
- **Proceed** — act only with clarity

REAL WORLD SCENARIO: THE TURNING POINT

A person realizes something is wrong.

Old Path

- panic
- react
- escalate

New Path

- pause
- document
- verify
- respond strategically

Result

Control is maintained.

The Hardest Truth in This Chapter

You cannot undo actions taken in emotion, but you can prevent them.

Final Strategic Shift

You do not give in to pressure. You manage the situation.

Final Close

When something goes wrong, your first instinct is to act. But your greatest protection is not action.

It is control.
Because the moment you react emotionally, you lose clarity.
When you lose clarity, you lose control.

But when you preserve first, you stay in position. From that position, you can win.

CHAPTER 10 TAKEAWAY

Don't react — record. Don't escalate — secure. Don't rush — control.

PART IV

SECTOR PLAYBOOKS

CHAPTER 11

Families, Elders, And Children

Building a Defensive Intelligence System at Home

Most people think protection is personal. Lock your account. Protect your phone. Watch your messages.

But here is the truth most people miss: You are only as secure as the least secure person connected to you.

The Family as a Target System

Attackers don't just target individuals.

They also target:

- families
- households
- relationships
- emotional bonds

Because families create:

- trust
- urgency

- emotional access

Emotional access is the easiest way in.

The Three Most Targeted Groups In Every Family, There Are Three Primary Points Of Vulnerability.

1. ELDERS

They are targeted because they:

- are trusting
- may not be familiar with new technology
- respond to authority
- value politeness and compliance
- may feel isolated

2. CHILDREN AND TEENS

They are targeted because they:

- are highly active online
- trust digital interactions
- share information freely
- respond to peer pressure
- do not recognize manipulation patterns

3. CAREGIVERS AND RESPONSIBLE FAMILY MEMBERS

They are targeted because they:

- act quickly to help
- carry responsibility
- respond emotionally to family needs
- may bypass verification under pressure

The Family Attack Model

Most family based manipulation follows this pattern:

- impersonate a family member
- create emotional urgency
- isolate the decisionmaker
- request immediate action

REAL WORLD SCENARIO: THE "GRANDPARENT CALL"

An older adult receives a call. "Grandma… it's me…"

The voice sounds familiar and distressed.

"I'm in trouble… please don't tell anyone."

Another voice comes on: "He needs help. This is urgent."

What Happens

- emotion overrides logic
- urgency removes verification
- isolation prevents consultation **Outcome (Without Defense)** Money is sent. Damage is done.

Outcome (With Defensive Intelligence)

- pause
- hang up
- call family directly
- confirm

Result

No loss.

The Family Defense Shift

Protection is no longer individual. It must be.

Intentional, Structured, And Communicated.

The G3TI Family Defense System

This becomes one of the most important systems in this entire book.

F.A.M.I.L.Y. Model

F — **Familiarize threats:** Educate every member on common tactics.

A — **Agree on protocols:** Set clear rules for communication and emergencies.

M — **Monitor information:** Be mindful of what is shared publicly.

I — **Implement verification:** No action without confirmation.

L — **Limit exposure:** Reduce unnecessary personal data online.

Y — **Yield to discipline:** Follow the system every time.

The Family Code Word System

Every family should have this.

What It Is

A simple, private word or phrase known only to trusted family members.

When It Is Used

In any situation involving:

- urgency
- distress
- requests for help
- unexpected communication

How It Works

If the person cannot provide the code word, do not proceed.

Why This Is Powerful

Because AI can clone voices, it cannot easily guess private, shared knowledge.

REAL WORLD APPLICATION

A message comes in: “I need help right now.”

Response

“What’s the code word?” **If They Can’t Answer** You stop.

Protecting Elders

What Elders Need

- simple rules
- clear instructions
- reassurance that it's okay to pause
- permission to question authority

Rules for Elders

- Never send money based on a phone call alone.
- Always call a trusted family member first.
- Do not act under pressure.
- Do not keep "secrets" for anyone who asks.

Important

You must make this clear to them: "You will never get in trouble for checking first."

Protecting Children and Teens

What They Face

- online impersonation
- social pressure
- gaming and chat manipulation
- exposure to strangers

What They Must Learn

- not everyone online is real.

- not every message is safe
- not every request is harmless

Rules for Children

- Do not share personal information.
- Do not engage with unknown contacts.
- Report anything that feels uncomfortable.
- Never act on pressure.

The Key

They must feel safe telling you anything, without fear.

Real World Scenario: The Teen Message

A teen receives: "Send this, or I'll expose you."

What Happens Without Training

- panic
- compliance
- secrecy

With Training

- report immediately
- do not respond
- preserve evidence

Result

Control is maintained.

The Family Communication Problem

Most families never talk about this. They assume everyone "knows what to do," but they don't.

The Family Defense Conversation

Every household needs a clear, structured conversation about:

- what to do
- what not to do
- who to call
- how to verify
- what emergencies look like
- what manipulation looks like

The Household Verification Protocol

Every family should adopt this rule:

No action is taken based on a single message, call, or request, ever.

The Shared Responsibility Principle

Protection is not the responsibility of one person. It is a shared system.

The Hardest Truth in This Chapter

You are not just protecting yourself. You are protecting the people who trust you, and those attackers believe they can use against you.

Final Strategic Shift

From: "I protect myself." To: "We protect each other."

Final Close

When a family operates with defensive intelligence, attackers lose their greatest advantage: emotional access.

Because a family that verifies, communicates, and supports each other is not a collection of targets. It is a unified defense system.

CHAPTER 11 TAKEAWAY

You are only as secure as the least protected person connected to you. Strengthen the family, and you strengthen every individual in it.

CHAPTER 12

Communities, Faith, And Public Trust

How Manipulation Spreads Through Groups and How to Stop It Before It Spreads

There is a level of manipulation that goes beyond the individual. It does not target just one person. It targets groups, communities, congregations, organizations, neighborhoods, and movements.

When trust is established within a group, it can be used to influence many people at once.

The Highest Value Target: Trust Networks

A trust network is any group of people who:

- believe in each other
- rely on shared values
- follow leadership
- communicate regularly
- act collectively

Examples

- churches and faith communities

- schools and parent networks
- nonprofits and advocacy groups
- neighborhood organizations
- professional associations
- online communities

Why These Groups Are Targeted

Because they already have what attackers need:

- built-in trust
- shared communication channels
- emotional alignment
- leadership structures

This means an attacker doesn't have to build trust from scratch. They only need to insert themselves into the system or impersonate someone within it.

The Community Manipulation Model

This is where things become more complex.

STAGE 1 — INFILTRATION OR IMPERSONATION

An attacker either:

- joins the group
- observes the group
- or impersonates someone within the group

STAGE 2 — TRUST ALIGNMENT

They:

- match language
- reflect values
- adopt tone
- appear consistent

STAGE 3 — MESSAGE INSERTION

They introduce:

- requests
- narratives
- instructions
- emotional appeals

STAGE 4 — TRUST LEVERAGE

Because the message appears to come from:

- a leader
- a member
- a trusted voice

It is accepted more easily.

STAGE 5 — SPREAD

Members:

- share the message
- act on it
- reinforce it

Result: One point of entry becomes many points of impact.

REAL WORLD SCENARIO:

THE FAITH COMMUNITY REQUEST

A message appears to be from a church leader: "Urgent need. Please assist immediately. Send support here."

What Happens

- members trust the source
- urgency is accepted, action is taken

What's Missed

No verification. **Outcome**

Loss affects multiple people.

The Power of Shared Trust

This is what makes community manipulation so dangerous. People do not question messages from:

- their pastor
- their school
- their leadership
- their internal network

Why

Because questioning feels like:

- distrust
- disrespect
- disloyalty

This is the exploit. Manipulation hides within respect.

The Emotional Multiplier Effect

Within communities, emotions amplify faster.

What Happens

One person feels urgency. They share it. Others feel it too. It spreads.

This Creates:

- panic
- rapid action
- reduced verification
- collective mistakes

REAL WORLD SCENARIO, THE COMMUNITY ALERT

A message spreads: "Warning: an immediate threat. Share this with everyone."

Without Control

- people react
- share quickly

- escalate fear

With Defensive Intelligence

- The message is verified.
- The source is confirmed.
- Communication is controlled.

Result

Spread is stopped.

The Leadership Responsibility

Leadership is not just influence. It is control over information flow.

If You Are a Leader

You must:

- verify before communicating
- avoid spreading unconfirmed information
- establish protocols
- train your people

Because one message from leadership can:

- calm a situation
- or create chaos

The G3TI Community Defense System

This becomes a national level concept.

T.R.U.S.T. Model

T — Test the message: Do not assume validity.

R — Route through verification: Confirm before sharing.

U — Understand the impact: What happens if this spreads?

S — Stop unverified flow: Do not pass along uncertainty.

T — Transmit with clarity: Communicate only confirmed information.

The "Do Not Forward" Rule

This is simple but powerful.

Rule

If it has not been verified, it is not shared.

Why

Because forwarding unverified information:

- multiplies risk
- spreads confusion
- creates damage

The Role of Faith Communities

Faith communities operate on:

- trust
- compassion
- service

- obedience to leadership

This makes them powerful, yet vulnerable.

Why

Because people are conditioned to:

- help quickly
- trust deeply
- act without hesitation

The Shift Required

From: "Help immediately." To: "Help responsibly."

Real World Scenario, The Help Request

A message appears: "A member is in urgent need. Please send assistance."

Defensive Response

- confirm through leadership
- verify the situation
- ensure legitimacy

Result

Real help is provided, and false requests are stopped.

The Community Verification Protocol

Every organization should have this.

Rules

- No urgent requests without confirmation.
- No financial action without dual verification.
- No message is trusted without source validation.
- No forwarding without clarity.

The “One Person Can Stop It” Principle

It only takes one person to:

- pause
- question
- verify

to stop an entire chain.

The Biggest Threat to Communities

It is not deception itself. It is unverified trust at scale.

The Strongest Truth in This Chapter

Trust is not the problem. Unverified trust is.

Final Shift

From: “If it came from us, it must be true.”

To: “If it is true, it must be verified.”

Final Close

Communities are built on trust. But trust without discipline can be turned against the very people it was meant to protect.

When one message is accepted without question, it can move through an entire group unchecked.

But when even one person pauses, verifies, and holds the line, the spread halts.

At that moment, the community is protected.

CHAPTER 12 TAKEAWAY

Trust is powerful, but only when it is verified before it is shared.

CHAPTER 13

Business, Government, And Organizational Defense

How to Protect Institutions, Employees, Operations, and Public Trust Against Manipulation Attacks

There was a time when organizations believed their greatest threat lay at the edge of the network, an outside attacker, a malware payload, a system breach, or a visible intrusion.

That model is no longer sufficient.

Many of the most damaging attacks against businesses, nonprofits, schools, municipalities, churches, and government institutions no longer begin with a firewall alert.

They begin with:

- a message
- a request
- a voice
- an email
- an assumption

- a false sense of urgency
- a trusted identity that is not what it seems

By the time the organization realizes what happened, the damage is already underway.

Money has moved. Information has been exposed. Credentials have been compromised. Public confidence has been shaken. Leadership is responding under pressure.

The institution is no longer operating from a position of strength; it is operating reactively.

That is why organizational defense must evolve.

The question is no longer only "Can our systems withstand attack?"

The question now is: "Can our people, our processes, and our leadership withstand manipulation?"

The Organizational Target

Organizations are high-value targets because they contain four things' manipulators want:

- money
- access
- authority
- trust

A business may hold financial assets, client records, vendor relationships, and brand credibility. A public agency may hold sensitive information, public authority, service continuity responsibilities, and decision-making power. A nonprofit may hold donor trust, mission credibility, and emotionally motivated support systems. A school may hold parent trust, student records, internal communications, and crisis sensitivity.

Each of these becomes a vulnerability when people inside the system are not trained to recognize manipulation.

One of the great misconceptions in modern defense is the belief that at-tackers only want to break in.

In many cases, they do not need to break in. They need to be let in, trusted in, or obeyed in.

And that changes everything.

The Institutional Attack Shift

Traditional security thinking focused on external intrusion.

Modern manipulation attacks often target:

- impersonating leadership
- exploiting workflow trust
- abusing urgency culture
- hijacking vendor relationships
- creating procedural bypass
- triggering emotionally driven decisions
- taking advantage of role-based authority
- using internal confusion against the institution

The modern attacker does not always fight your controls head-on. They move around them by targeting human behavior.

A finance manager who believes the request is legitimate. A staff member afraid to question authority. A supervisor who acts quickly to avoid delay. A team member who assumes someone else has verified the message. An executive who responds under time pressure without process discipline. This is the real institutional vulnerability:

When trust outruns verification.

The Cost of Manipulation Inside an Organization

Consequences are rarely minor.

They can include:

- fraudulent wire transfers
- payroll diversion
- vendor payment rerouting
- credential theft
- unauthorized system access
- exposure of client, citizen, or employee information
- legal liability
- regulatory scrutiny
- media exposure
- contract loss
- operational downtime

- leadership credibility damage
- erosion of internal morale
- public distrust

And the most dangerous part: Many failures begin as something that seems routine.

Organizations must stop treating manipulation as a cyber issue and start treating it as an enterprise-wide operational risk.

The Organizational Manipulation Chain

Most successful attacks follow a repeatable attack chain.

1. RECONNAISSANCE

The attacker studies the organization and learns:

- leadership names
- department roles
- vendor relationships
- public facing language
- organizational structure
- payment processes
- internal tone
- timing patterns
- public calendars
- press releases
- job postings

- social media habits

Organizations often reveal far more than they realize.

The attacker is not starting blind; they are building context.

2. TARGET IDENTIFICATION

Specific people are identified, usually those with:

- authority to approve
- access to sensitive systems
- responsibility to move quickly
- customer facing trust
- financial control
- executive proximity
- communications authority
- operational importance

Often not the most senior person, but the most accessible one.

3. PRETEXT CREATION

A believable scenario is presented:

- urgent payment correction
- account verification
- contract update
- travel related transfer
- executive request

- legal threat
- security concern
- employee update

The attacker creates a pretext for contact. This is where the attack takes shape.

4. LEGITIMACY SIMULATION

The attacker mimics:

- tone
- timing
- formatting
- internal language
- leadership style
- vendor identity

The message is no longer evaluated on its own; it is evaluated through assumed trust.

5. PRESSURE INTRODUCTION

Pressure appears, either directly:

- "Need this immediately."
- "This cannot wait."
- "Handle this now." **Or subtly:**
- "I'm in a meeting; just take care of it."
- "This is sensitive. Don't involve others yet."

The function of pressure is always the same: to reduce the time for thought, consultation, and verification.

6. PROCESS BYPASS

This is the institutional breaking point.

Manipulation succeeds when a person abandons process to relieve immediate pressure:

- bypassing callback verification
- skipping secondary approval
- changing payment details without confirmation
- sending confidential records without authorization
- accepting a voice as proof of identity
- acting on a message without validating the domain
- assuming someone else already verified it

Once the process is bypassed, the institution is no longer protected by its design. It is running on human improvisation under pressure.

7. EXTRACTION, ACCESS, OR IMPACT

The attacker gets what they came for:

- funds
- credentials
- information
- internal visibility

- procedural compliance
- public confusion
- operational advantage

By the time the institution realizes what happened, the damage has already crossed into consequences.

The Organizational Weaknesses Attackers Count On

Attackers rely on:

- urgency culture
- authority pressure
- process fatigue
- verification stigma
- fragmented communication
- role isolation
- overconfidence
- public information leakage
- lack of escalation culture
- no tabletop practice

These weaknesses create the ideal environment for manipulation.

Why Leadership Sets the Defense Culture

Defense is not just policy; it is leadership. **If leadership:**

- When speed is prioritized, staff will rush
- reacts poorly to questions; staff will stop verifying

- bypasses rules, staff will assume rules are optional
- communicates vaguely, confusion spreads

A defensible organization is built when leadership makes this clear.

Verification is not a disrespect. Verification is protection.

The G3TI Organizational Defense Doctrine

A strong core system:

S.H.I.E.L.D. Model

S — Standardize critical workflows: Sensitive actions must follow a documented, repeatable process.

H — Harden verification points: No money, data, or access moves without independent validation.

I — Interrupt urgency culture: Speed never outranks accuracy in high-risk actions.

E — Elevate suspicious activity early: Staff must know when to report, pause, and escalate.

L — Limit single point authority: No one person controls an entire sensitive transaction.

D — Drill the response: Training must move from awareness to repetition.

This is not just a framework; it is an operating philosophy.

Critical Control Areas

1. FINANCE AND PAYMENTS

Finance is one of the most targeted functions. Required safeguards:

- dual approval
- callback verification
- no banking changes from email alone
- documented confirmation
- exception logging
- waiting periods for new payment destinations
- escalation for unusual requests

Core rule: No financial instruction is acted on because it "looks right." Only after it has been fully verified.

2. EXECUTIVE IMPERSONATION

The higher the authority, the greater the leverage. Required safeguards:

- executive verification protocol
- known callback numbers
- no approvals by email/text alone for high-risk matters
- voice verification plus secondary confirmation
- leadership discipline in communication
- explicit permission for staff to verify Leadership discipline is a security control.

3. HUMAN RESOURCES AND PAYROLL

HR holds sensitive identity and financial information.

Required safeguards:

- identity verification before payroll changes
- no direct deposit changes from email alone
- out of band confirmation
- scrutiny for "urgent employee issues"
- logs and approvals for sensitive data
- training for HR specific social engineering

4. IT HELP DESK AND ACCESS CONTROL

Help desks are targeted because they resolve problems quickly.
Required safeguards:

- strong identity verification
- no resets based on emotional pressure
- no privilege changes without documentation
- incident logging
- secondary approval for highprivilege changes
- repeated social engineering drills

5. PUBLIC COMMUNICATIONS AND REPUTATION

Institutions can be manipulated into spreading falsehoods. **Required safeguards:**

- verified source policy
- approved communication chain
- pause and confirm protocol
- clear distinction between confirmed and unconfirmed
- internal coordination before external messaging

6. NONPROFITS AND MISSION DRIVEN ORGANIZATIONS

Mission-driven groups are vulnerable because they act quickly to help. **Required safeguards:**

- donor verification
- approval chain for emergency appeals
- official only donation channels
- awareness of impersonation risk
- no changes to donation instructions without validation

7. PUBLIC AGENCIES AND GOVERNMENT OPERATIONS

Manipulation can undermine public confidence and service continuity.

Required safeguards:

- role-based verification
- interagency callback procedures
- procurement integrity checks
- suspicious instruction reporting
- documented emergency communication chain
- verification even under pressure (except life safety emergencies)

Building the Defensive Organization

A defensible organization requires:

- **Policy** — document the rules
- **Procedure** — define the steps
- **Training** — teach the why and how
- **Rehearsal** — practice under realistic conditions
- **Escalation** — make reporting simple
- **Leadership Modeling** — leaders follow the rules
- **Review** — analyze incidents to improve

Without this cycle, institutions rely on memory and luck rather than defense.

The Organizational Response to a Suspected Manipulation Event

- Pause the action
- Preserve evidence
- Contain exposure
- Verify independently
- Escalate internally
- Assess impact
- Act proportionately
- Learn and strengthen Structure is what brings back control.

The Role of Culture in Organizational Survival

A defensible culture ensures:

- People are praised for checking.
- No one is shamed for slowing down.
- Pressure is recognized as a warning sign.
- Executives expect verification.
- Suspicious activity is elevated early.
- Urgency is not mistaken for legitimacy Culture answers the silent question:

"Am I allowed to stop this?"

The answer must always be yes.

The Most Important Training Shift

Organizations must train people to recognize:

- executive impersonation
- payroll diversion
- vendor fraud
- voice based manipulation
- urgency abuse
- authority pressure
- process bypass attempts
- emotional coercion
- misinformation spread
- internal confusion exploitation

This is not cyber awareness. It is defensive operational intelligence.

The G3TI Organizational Command Principle

One principle strong enough to govern an entire institution.

No sensitive action moves on urgency alone.

This applies to:

- money
- access
- records

- crisis communications
- confidential disclosures
- leadership directives outside normal process

REAL WORLD SCENARIO:

CITY DEPARTMENT PRESSURE EVENT

A department employee receives what appears to be an urgent payment request from a senior official.

In a weak organization, the employee acts. In a defensible organization, the employee pauses.

They follow protocol:

- independent verification
- secondary approver
- callback using known official number
- documentation
- notification to internal control

The request is flagged as false before damage occurs.

This is organizational maturity: not perfection, but interruption.

The Hardest Truth in This Chapter

Institutions rarely fail because they lack policies. They fail because, in the critical moment:

- the policy was not followed.
- verification was not normalized

- urgency overran discipline
- leadership culture discouraged hesitation
- staff did not know they had permission to stop

This is not about bureaucracy. It is about institutional survival under pressure.

Final Strategic Shift

Manipulation is not a side issue.

It is a direct threat to:

- continuity
- finance
- public confidence
- mission execution
- leadership credibility
- operational integrity

The strongest organizations are not those that trust blindly but those that equip their people to question, verify, pause, escalate, and protect the mission under pressure.

Final Close

Institutions do not break only when systems fail. They break when people are pressured into circumventing the controls designed to protect them.

A manipulated employee can move money. A manipulated leader can sow confusion. A manipulated institution can damage its own credibility faster than any outside attacker.

Defense in this era is not only technical. It is behavioral, procedural, cultural, and leadership-driven.

When an organization embeds discipline in the bloodstream of its operations, manipulation ceases to be a silent threat; it becomes an interrupted one.

CHAPTER 13 TAKEAWAY

A secure organization is not one that never gets targeted. It is one that refuses to let urgency outpace verification.

CHAPTER 14

Law enforcement, fusion, and protective intelligence

How AI Enabled Manipulation, Synthetic Deception, and Influence Attacks Now Intersect with Public Safety, Victim Protection, and Intelligence Led Prevention

For years, law enforcement and public safety agencies have been trained to recognize danger in familiar locations.

- a weapon
- a suspect
- a threat call
- a fraud complaint
- a domestic incident
- a burglary pattern
- a narcotics network
- a criminal enterprise
- a physical crime scene

Those threats still exist. However, a new category of threat is spreading across every jurisdiction, community, demographic group, and professional sector.

Manipulation Of Human Perception For Criminal, Coercive, Financial, Reputational, And Operational Advantages.

This is the new battlespace, and many agencies are not yet organized to see it clearly.

Victims don't walk around saying, "I was targeted by synthetic influence and AI-assisted behavioral manipulation."

Instead, they say:

- "I received a call from my son and sent money."
- "My boss told me to transfer funds."
- "Someone threatened to expose me."
- "A message spread throughout the church."
- "We thought the warning was real."
- "The employee followed instructions."
- "The family panicked."
- "The community reacted."

Beneath these cases lies a deeper truth: a person's decision-making was shaped by pressure.

That is not a side issue. It is a challenge to protective intelligence.

The Public Safety Consequences of Synthetic Manipulation

Public safety has traditionally focused on incidents after they occur:

- after the assault
- after the fraud
- after the disturbance
- after the loss
- after the threat is acted upon

But manipulation-driven harm often begins before any visible crime scene appears.

It begins in:

- a message thread
- a cloned voice call
- a fake authority contact
- a synthetic relationship
- a false narrative
- an engineered emergency
- a victim under pressure
- a patrol call shaped by misinformation
- a household destabilized by fear

The initial crime scene is often psychological and informational before it becomes physical or reportable.

That shifts the way law enforcement and security systems need to approach things.

Why This Is a Law Enforcement Issue: Not Just a "Tech Issue

It's a dangerous mistake to say:

- "That's for cyber."
- "That's for fraud."
- "That's for the bank."
- "That's not patrol."
- "That's not victim services."
- "That's not fusion." This mindset is overly narrow. Manipulation now impacts:
- patrol responses
- victim behavior
- witness reliability
- escalation patterns
- extortion
- coercive control
- domestic violence lethality
- financial crimes
- elder exploitation
- juvenile exploitation
- false reporting
- misinformation spread
- public confidence

- officer safety
- community panic
- evidence handling
- interagency trust

This is no longer a single-unit problem; it is a cross-functional threat vector.

The Evolution of the Offender Model

Traditional offenders:

- steal
- assault
- trespass
- threaten
- physically approach victims

Modern manipulators might never be physically present. They could operate from another city, state, or country using:

- cloned voices
- synthetic video
- fraudulent authority
- romance based trust building
- blackmail dynamics
- emergency narrative injection
- impersonation of institutions

- AI assisted conversation scaling

They may never touch the victim, yet they can still:

- evoke an emotional response from them
- isolate them mentally
- redirect funds
- trigger panic
- cause confrontation
- influence behavior
- obstruct judgment
- increase community harm

The crime is increasingly mediated by engineered human reactions. Public safety should shift from asking only, "Who committed the act?" to also considering: "How was the victim manipulated into exposure?"

The Protective Intelligence Lens

Protective intelligence is more than knowing that threats are present.

It involves recognizing:

- pre-incident indicators
- behavioral escalation signals
- vulnerability intersections
- attack pathways
- pressure signatures

- influence patterns
- decision points before harm occurs

Modern manipulation should be part of a protective intelligence framework because it is:

Deliberately shaping the human behavior to produce exploitable outcomes.

This is not just a collection of scam stories. It is a systematic approach to threats.

The Four Public Safety Domains Impacted by Modern Manipulation

- **Victimization** — people are coerced into harmful decisions.
- **Incident escalation** — false narratives fuel conflict, panic, or retaliation.
- **Institutional disruption** — agencies and organizations are being manipulated operationally.
- **Trust deteriorates** — communities lose confidence in what is true.

When trust declines, public safety declines as well.

Patrol and First Response Realities

Patrol officers are often the first to encounter victims whose judgment has already been manipulated.

A surface-level report may obscure a deeper influence pathway.

Examples include:

- blackmail
- threats
- family panic
- domestic conflict triggered by misinformation
- extortion involving explicit images
- authority impersonation
- elder confusion
- business related fraud Patrol must increasingly understand not only the criminal act but also the influence pathway that led to it.

What Patrol Must Be Trained to Ask

Not just:

- What happened?
- Who called?
- What was taken?

But also:

- How were you contacted?
- What pressure was used?
- Was urgency present?
- Was authority claimed?
- Were you isolated?
- What were you afraid would happen?

- Did you preserve the communication?
- Is this part of a broader pattern?
- Is there risk of panic, retaliation, or self-harm now? These questions transform a routine report into a threat picture.

Detectives, Investigations, and the New Evidentiary Landscape

Old assumptions no longer hold:

- a voice recording is not automatically proof
- a screenshot may not tell the whole story
- a video may be synthetic
- a text thread may be part of a larger influence sequence Investigators must assess:
- channel sequencing
- timing of pressure
- narrative construction
- grooming indicators
- metadata
- account origination
- social engineering architecture

Without understanding manipulation, investigators may misinterpret victims as either careless or complicit.

With understanding, the behavior becomes intelligible, and the case becomes clearer.

Victim Services and Trauma

Informed Response

Victims of manipulation are often harmed twice.

1. by the offender
2. by systems that misunderstand them Victims may present with:

- embarrassment
- panic
- self-blame
- fear of judgment
- reluctance to report
- suicidal thoughts in severe extortion
- emotional collapse

Victim services must understand:

Manipulation disables the very judgment that people later blame themselves for losing.

The correct approach asks:

- What pressure were you under?
- What made this feel real?
- What were you afraid would happen?
- What do you need to feel safe right now?

Domestic Violence, Coercive Control, and Digital Manipulation

Manipulation overlaps significantly with coercive control.

Offenders may use:

- impersonation
- fake messages
- spoofed calls
- synthetic evidence
- threats of exposure
- narrative distortion
- digital monitoring
- emotional destabilization

This can increase the risk of lethality. Protective intelligence must ask:

- Is the offender manipulating perception?
- Is synthetic communication involved?
- Is the victim being isolated digitally?
- Is the support system being targeted?

Elder Exploitation as a Protective Intelligence Priority

Elders are highly targeted because they may:

- trust authority
- avoid conflict
- fear embarrassment
- be isolated

- have accessible funds

Agencies must treat elder exploitation as:

- emotional exploitation
- authority exploitation
- trust exploitation

Not just financial loss.

Early intervention is critical because repeated targeting is common.

Juveniles, Teens, and Digital Extortion Environments

Youth face a threatening environment shaped by the following:

- secrecy
- image culture
- peer pressure
- online normalization

They may be manipulated through:

- fake peers
- sexual extortion
- threats of exposure
- gaming platform contact
- AI enhanced impersonation

Youth often fear humiliation more than the offender does.

Adults must watch for:

- sudden panic

- withdrawal
- obsessive phone distress
- secrecy
- unusual compliance
- mentions of exposure threats

Fusion, Analytic Environments, and Threat Integration Manipulation intersects with:

- fraud
- cyber
- public safety calls
- elder exploitation
- domestic violence
- youth victimization
- misinformation
- organized crime

Fusion centers can add value by:

- identifying recurring pretexts
- tracking impersonation themes
- integrating cyber indicators
- producing bulletins
- connecting local incidents to broader patterns

REAL TIME CRIME CENTERS AND THE INFORMATION BATTLESPACE

RTCCs must recognize:

- Call narratives may be false.
- Social media signals may be synthetic.
- Community messages may create false urgency.
- Victims may act based on engineered panic

RTCCs must ask:

- Is the information itself reliable?
- Has the event been manipulated?
- Is the reported urgency real or engineered?

The Investigative Question That Changes Everything

What was the victim, witness, employee, family member, or institution led to believe, and by whom?

This question reveals:

- motive
- pressure
- method
- timing
- vulnerability
- intended outcome

The G3TI Public Safety Manipulation Model

T.R.I.A.D. Model

T — **Threat narrative:** What story was inserted?

R — **Response pressure:** What emotion or urgency was used?

I — **Influence pathway:** How was the target/person moved?

A — **Action demand:** What were they pushed to do?

D — **Defensive interruption:** Where could it have been stopped? This model makes manipulation operationally visible.

Officer Safety and Authority Impersonation

Officers can also be targeted through:

- fake command directives
- spoofed interagency contacts
- manipulated evidence narratives
- false emergency notifications

A message that invokes authority is not automatically authentic. Verification must be performed internally.

Command Staff, Supervisors, and Agency Culture

If supervisors normalize rushed, informal directives, the agency becomes vulnerable to such directives.

If staff are punished for verifying, they will comply with manipulation. The correct model:

Move when necessary. Verify whenever possible. Protect the mission from false urgency.

What a Modern Agency Training Program Should Include Patrol

- recognizing influence based victimization
- preserving digital evidence
- calming manipulated victims

Detectives

- synthetic deception awareness
- influence pathway reconstruction

Victim Services

- trauma informed stabilization
- shame reduction

Supervisors

- escalation pathways
- communication discipline

Analysts / Fusion Personnel

- pattern recognition
- pretext tracking

Public Information

- rumor control
- message verification

The Public Safety Response Shift

From: "Did a crime happen?"

To:

- What pressure was applied?
- How was trust exploited?
- What decision was engineered?
- What is the repeat risk?
- What can be interrupted before the next victim?

The Hardest Truth in This Chapter

Victims will continue to be misunderstood unless agencies learn to distinguish between the following:

bad judgment and manipulated judgment. Those are not the same.

Final Strategic Shift

A person can be robbed without a weapon. A family can be destabilized without a suspect entering the home. A victim can be controlled without being in close proximity. An agency can be manipulated without a network breach. A community can be moved without a visible offender.

Perception itself has become a threat surface.

Final Close

The future of public safety will be secured by agencies that understand the **following:**

- how modern harm is engineered before it becomes visible.
- how victims are maneuvered before being blamed
- how trust is weaponized before institutions recognize it.

The threat is no longer only what a criminal can do with his hands. It is what he can make others do with their own fear, confusion, urgency, and belief.

When public safety recognizes that it no longer arrives late at crime scenes, it begins to intervene earlier in the process.

CHAPTER 14 TAKEAWAY

In the AI era, public safety must protect not only people and places but also perception, trust, and decision-making under pressure.

PART V

RECOVERY AND RESPONSE

CHAPTER 15

The Second Mindset

How to Rebuild, Reposition, and Become Harder to Manipulate After Exposure

There are two types of people after an incident. Those who say: "I hope this never happens again."

And those who say, "This will never happen to me the same way again." This chapter is intended for the second person.

Recovery Is Not Enough

Stabilization is not enough. Even understanding what happened is inadequate.

If nothing changes, the same pattern will repeat under different circumstances.

The First Truth of This Chapter

You cannot return to who you were, and you should not try.

Because the person you were before lacked the awareness you now have.

The Shift

From: "I want things to go back to normal." To: "I operate at a higher level now."

The Second Mindset

This is what separates:

- repeat victims from
- resilient operators

The Second Mindset Is Built on Three Things

1. Clarity
2. Control
3. Consistency

Clarity —You See Differently Now

After exposure

You begin to notice:

- pressure patterns
- emotional triggers
- manipulation language
- timing tactics
- authority positioning

What used to feel normal now feels different. Noticeably so. Off. This is not paranoia. This is awareness.

Control — You Move Differently Now

Before, you may have:

- responded quickly
- trusted easily
- acted under pressure

Now, you:

- pause automatically
- verify consistently
- slow down intentionally

You are no longer reactive. You are in control.

Consistency, You Do Not Break Your Own Rules

This is where most people fail.

They say, "I'll be more careful next time," but under pressure, they do:

- rush
- assume
- skip steps

The second mindset does not do this. It follows the process every time, without exception.

The G3TI Second Mindset Framework

R.E.S.E.T. Model

R — Recognize patterns: You've seen this before.

E — **Establish control:** You do not rush.

S — **Slow everything down:** Time is your advantage.

E — **Execute verification:** No action without proof.

T — **Trust your process:** Not your impulse.

The Danger of Partial Learning

One of the biggest risks after an incident is this:

You learn just enough to regain your confidence.

This is a problem because confidence without discipline leads to repeated exposure to risk.

REAL WORLD SCENARIO, THE "I GOT THIS NOW" TRAP

A person is targeted again. They think, "I've seen this before," and move quickly, skipping verification.

Result: Different scenario. Same outcome.

Lesson: Awareness without discipline is not protection.

The Identity Shift

This is where transformation happens.

Before: "I try to be careful." **After:** "I am not easily moved." **This is who you become:**

- controlled under pressure
- calm under urgency

- disciplined under emotion
- strategic under uncertainty

The "Pause as Default" Principle

Before, pausing was something you had to remember. Now it is automatic. This is powerful because most manipulation relies on speed.

Remove speed, and you lose your advantage.

REAL WORLD SCENARIO, THE REPEAT ATTEMPT

A message arrives: "Urgent. Immediate action required."

Old response: Engage.

New response: Pause. Assess. Verify.

Result: No impact.

The Trust Reconstruction Problem

After an incident, many people go too far in one direction. Either they trust everything again or they trust nothing at all. Both are problems.

The correct position: You trust after verification.

This is disciplined trust. Neither blind nor broken.

The Environmental Upgrade

Your mindset alone is not enough. Your environment must change as well. **You must:**

- reduce exposure

- strengthen verification systems
- communicate with trusted people
- implement safeguards

The people around you must also adjust. Your environment shapes your decisions.

If Others Are Rushing You

You must hold your position.

The Power of Saying No

You must be able to say:

- "I need to verify that."
- "I'm not comfortable acting right now."
- "I will get back to you." Without explanation.

REAL WORLD SCENARIO, THE PRESSURE MOMENT

"Just handle it. It's urgent."

Second mindset response: "I'll verify and follow up."

Result: Control remains.

The Reputation Shift

You may worry, "What will people think?"

The truth: People respect control, not reaction.

The Strongest Truth in This Chapter

You are not rebuilding to avoid fear. You are rebuilding to operate at a higher level.

Final Shift

From: "I hope this doesn't happen again."

To: "I am not the same person anymore."

Final Close

Exposure is not the end of your strength. It is the beginning of your awareness.

Once you see how manipulation works, you can't unsee it.

Once you learn to control your response, you cannot be moved the same way again.

Not because the world changed, but because you changed.

CHAPTER 15 TAKEAWAY

You don't go back, you level up.

CHAPTER 16

The Psychology Of Aftermath

Shame, Anger, Revenge, Identity, and the Battle to Reclaim Your Mind After Manipulation

There is a second impact that occurs after every manipulation event.

It is neither financial, digital, nor procedural.

It is psychological.

For many people, it is more damaging than the original loss itself.

The Moment After Realization

Everything slows down. You replay it.

- the message
- the call
- the decision
- the moment you acted And one thought begins to repeat:

"How did I let this happen?"

This is where the real battle begins. The attacker is gone, but the impact remains. Now you are dealing with yourself.

The Five Psychological Responses

Almost everyone goes through some version of these. Not because they are weak, but because they are human.

1. SHAME — "I SHOULD HAVE KNOWN"

Shame is quiet.

It doesn't shout. It sits inside, repeating:

- "That was obvious."
- "I'm smarter than this."
- "I can't believe I did that."
- "I can't tell anyone."

What Shame Does

- isolates you
- silences you
- delays recovery
- prevents reporting
- protects the attacker

The Truth About Shame

Shame is not a reflection of your intelligence. It is a response to violated trust.

2. ANGER — "SOMEONE IS GOING TO PAY FOR THIS"

Anger comes next, directed at:

- the attacker
- the situation
- yourself
- sometimes others

What Anger Does

- pushes you to act fast
- encourages confrontation
- creates risk
- leads to impulsive decisions

The Danger

Anger feels powerful. But it removes control.

3. SELF-BLAME — "THIS IS MY FAULT"

One of the most destructive responses. **People say:**

- "I should have checked."
- "I rushed."
- "I ignored the signs."

What This Does

- weakens confidence

- creates hesitation
- leads to withdrawal

The Reality

Yes, a decision was made, but it was made under engineered pressure.

4. REVENGE — "I'M GOING TO FIX THIS MYSELF "

A critical danger point.

What This Looks Like

- trying to outsmart the attacker
- engaging further
- seeking immediate recovery
- responding aggressively

The Problem

This leads to:

- further loss
- escalation
- exposure

The Truth

You do not regain control through revenge. You regain it through discipline.

IDENTITY DISRUPTION — "THIS IS NOT WHO I AM"

This is deeper.

Your self-image is shaken:

- "I'm careful."
- "I'm smart."
- "I don't make mistakes like this."

This Creates

- confusion
- loss of confidence
- hesitation
- overcorrection

The Internal Loop

This is where everything repeats:

- replay the event
- feel shame
- feel anger
- blame yourself
- think about revenge
- repeat

This loop can last days, weeks, or months. If left uninterrupted, it becomes identity.

The G3TI Psychological Recovery Model

One of the most important frameworks in the entire book.

R.E.C.L.A.I.M. Model

R — **Recognize the impact:** This affected you, and that's real.

E — **Eliminate selfdestruction:** Stop blame, revenge, and panic cycles.

C — **Create mental distance:** Step out of the loop.

L — **Label the manipulation:** Understand what was done to you.

A — **Accept the event:** Without defining yourself by it.

I — **Implement control:** Apply systems moving forward.

M — **Move forward deliberately:** Not emotionally.

The Shift From Emotion to Control

You must move from: "I feel this strongly…" To: "I understand what happened."

REAL WORLD SCENARIO: THE BREAKDOWN

A person sits alone after the incident. They feel:

- shame
- anger
- regret

PATH 1 — STAY IN THE LOOP

- withdraw
- replay
- isolate

PATH 2 — INTERRUPT

- acknowledge
- step back
- analyze
- stabilize

Result: Control returns.

The Role of Understanding

When you understand:

The tactics

- the pressure
- the structure

You stop asking: "What's wrong with me?"

And start asking, "What was done here?"

The Power of Reframing

Old frame: "I failed."

New frame: "I was targeted, now I'm trained."

The Emotional Discipline Rule

You are allowed to feel. You are not allowed to act on uncontrolled emotion.

The "No Decisions in Distress" Rule

When you are in this state, you do not do:

- confront
- respond
- decide
- escalate

You wait until clarity returns.

REAL WORLD SCENARIO, THE REVENGE MOMENT

A person considers reaching out to the attacker again. **Old response:**

Engage.

New response:

Stop.

Result:

No further damage.

The Recovery Identity

This is where transformation happens. **You become:**

- more aware
- more disciplined

- more controlled
- more observant You do not become:
- fearful
- withdrawn
- broken

The Strongest Truth in This Chapter

You were not defeated. You were exposed. Exposure can be turned into strength.

Final Shift

From: "I can't believe this happened."

To: "I understand what happened. I've adjusted."

Final Close

The greatest damage is not what was taken. It is what you believe about yourself afterward.

If you allow shame to define you, you lose twice. But if you understand what happened… if you regain control… if you rebuild with discipline… Then the event does not weaken you. It sharpens you.

Once you are sharpened, you cannot be moved the same way again.

CHAPTER 16 TAKEAWAY

You are not defined by the moment you were manipulated, you are defined by how you rebuild afterward.

CHAPTER 17

Building Personal Defensive Intelligence to Live, Operate, and Think at a Higher Level Every Day

There is a difference between knowing what manipulation is... and becoming the kind of person who is hard to manipulate.

A person can read every warning, learn every tactic, and understand every framework, and still fail in the moment if their way of life has not changed. This chapter is not about information. It is about transformation.

The strongest defense is not found in a single good decision. It is found in a disciplined pattern of thinking so consistent that pressure no longer controls your movements.

That is what personal defensive intelligence means.

It is not a trick. It is not a slogan. It is not a temporary alertness. **It is a way of life in which:**

- Your judgment is protected by structure.
- Your emotions are governed by discipline.

- Your trust is verified through verification
- Your actions are filtered through thought.

This is how you stop being reactive in a manipulative age. This is how you become stable in a world designed to destabilize people.

What Personal Defensive Intelligence Really Is

Personal defensive intelligence is the daily capacity to:

- notice pressure without surrendering to it
- separate emotion from decision
- verify before acting
- recognize manipulation patterns early
- protect attention, identity, time, money, and relationships
- remain calm in high-pressure moments
- build sys
- terms that reduce unnecessary vulnerability
- think beyond the moment and act with consequence in mind Manipulation rarely wins by brilliance. It wins by inconsistency.

It wins when you are disciplined one day and careless the next. Cautious with strangers, reckless with familiarity. Clear when calm, compromised when tired, stressed, lonely, proud, distracted, offended, or rushed.

The issue is not only what is happening around you. It is: What kind of operator are you becoming?

The Hard Truth Most People Avoid

Many people do not have a manipulation problem. They have a consistency issue.

They know better, but they do not live better.
They understand the need to slow down but still rush. They understand the need to verify but still assume. They understand the need for emotional control but still let the moment decide. They understand the need for boundaries, but still overexpose themselves to pressure, noise, distraction, and urgency.

Personal defensive intelligence cannot be occasional. It must become operational, a personal operating system.

The Daily Battle for Your Decision-Making

Manipulation does not begin at the moment of crisis.

It begins in daily habits that weaken judgment:

- constant distraction
- emotional impulsivity
- information overload
- digital overexposure
- poor boundaries
- reactive communication
- ego driven responses
- fatigue based decisions
- dependence on speed

- discomfort with pausing
- unwillingness to verify
- intolerance of uncertainty

If you live in a constant state of urgency and overstimulation, the attacker does not need to create instability. It already exists.

Building personal defensive intelligence is not only about defending against external threats. It is about creating an internal environment where clarity endures pressure.

The G3TI Personal Defensive Intelligence Model

A life level framework:

P.R.E.P.A.R.E. Model

P — **Protect** your attention

R — **Regulate** your emotions

E — **Evaluate** before acting

P — **Pause** under pressure

A — **Authenticate** trust

R — **Reduce** unnecessary exposure

E — **Execute** with discipline

This is not a chapter framework. It is a life framework.

1. PROTECT YOUR ATTENTION

Your attention is the gateway to your mind.

Whatever controls your attention shapes your thoughts. Whatever shapes your thoughts influences your emotions. Whatever influences your emotions affects your decisions. Whatever affects your decisions shapes your outcome.

Manipulators understand this.

That is why they use:

- shock
- urgency
- drama
- novelty
- conflict
- fear
- attraction
- outrage
- false emergencies
- emotional hooks

A defensive person learns:

Not everything that demands your attention deserves it.

Daily attention discipline:

- turn off nonessential notifications
- schedule communication review
- avoid decisions while multitasking

- do not let emotional messages dictate your pace
- build silence into your day
- let messages wait while your mind stays clear

A person who cannot sustain attention cannot reliably maintain control.

2. REGULATE YOUR EMOTIONS

Emotion is not a weakness. Unmanaged emotion is vulnerability. **A person who cannot regulate:**

- fear will be rushed
- anger will be baited
- loneliness will be exploited
- pride will resist checking
- shame will stay silent

Emotional regulation is a protective skill.

You must learn to identify emotional activation in the moment, not afterward.

Signals include:

- rising heart rate
- pressure in the chest
- racing thoughts
- urge to respond immediately
- fear of waiting

- need to defend yourself quickly These are signals, not instructions.
- never respond immediately to high-emotion messages
- avoid major decisions when feeling angry, afraid, ashamed, or euphoric
- write before you send
- say your emotions out loud before acting
- ask, "Is this emotion informing me or driving me?"
- normalize stepping back

The goal is not numbness but governance.

3. EVALUATE BEFORE ACTING

Most people live in reactive mode. A defensively intelligent person inserts evaluation between an event and an action.

Ask:

- What is actually happening?
- What is being asked of me?
- What do I know for sure?
- What do I not know yet?
- What is the pressure mechanism?
- What happens if I wait?
- Who benefits from my speed?

- What would this look like if it were deceptive? **Daily evaluation habits:**
- review context before responding
- read twice before replying
- separate facts from assumptions
- ask follow up questions
- never let familiarity replace analysis
- test the situation before trusting the appearance

Evaluation protects against manipulation, conflict, misjudgment, and false urgency.

4 PAUSE UNDER PRESSURE

The pause is not passive. It is an exercise of strategic power. **The pause means:**

- You do not surrender your timing.
- You do not let urgency dictate your actions
- You do not confuse speed with strength.
- you do not let pressure collapse your options. Whoever controls the tempo often determines the outcome. Daily pause practices:
- count to ten before answering pressure-based questions
- wait before sending emotional messages
- create a nonresponse window for conflict
- step away physically if needed

- refuse to act in the first wave of emotion
- say, "I'll review this and get back to you"

The pause is under control.

5. AUTHENTICATE TRUST

Trust is necessary, but it must be structured. **Disciplined trust means:**

- trust with verification
- trust with boundaries
- trust with standards

You no longer say, "This feels trustworthy."

You say, "This might be trustworthy, but I will confirm."

Daily trust disciplines:

- Do not confuse familiarity with legitimacy.
- Do not confuse confidence with credibility
- Do not confuse urgency with authenticity
- Verify identity before acting
- Create standards for what requires proof

Disciplined trust keeps you open without leaving you exposed.

Reduce Unnecessary Exposure

Some of the strongest defense occurs before contact. **Ask:**

- What am I making too easy to access?

- What am I giving away without thinking?
- What systems are too loose?
- What habits increase my risk?

Digital exposure: oversharing, weak privacy, public routines. Emotional exposure: telling too much too early. Procedural exposure: acting alone on sensitive decisions. Social exposure: weak boundaries, difficulty saying no. Not every vulnerability should be addressed in the moment. Many should be mitigated in advance.

Execute With Discipline

Knowledge without action is decoration. Awareness without discipline is fragile.

Execution means:

- You actually verify
- You actually pause
- You actually preserve evidence
- You actually decline pressure
- You actually follow your rules

Discipline matters more than inspiration. Inspiration is strong in calm moments, while discipline is strong in stressful ones.

Daily execution habits:

- build default protocols
- rehearse responses to urgency
- keep written personal rules

- avoid exceptions because something feels "probably fine".
- let process be stronger than mood

The Personal Rules That Change Everything

Examples of nonnegotiable:

- I do not send money under pressure.
- I do not verify within the same channel.
- I do not respond when emotionally activated.
- I do not make major decisions when fatigued or ashamed.
- I do not share private information to resolve urgent requests.
- I do not let title or tone replace evidence.
- I do not let my ego stop me from checking.
- I do not confuse kindness with accessibility.

These rules protect your future self from your present self under pressure.

Building a Defensive Life, Not Just a Defensive Moment

A defensive life includes:

Relationally: You choose people carefully and value consistency.

Digitally: You secure accounts and reduce oversharing.

Financially: You establish approval habits and waiting periods. **Emotionally:** You strengthen your ability to tolerate discomfort. **Professionally:** You reject urgency culture.

Spiritually and mentally: you build inner stability.

This is not fear.

It is order.

The Enemy of Personal Defensive Intelligence

The real enemies are:

- ego
- fatigue
- distraction
- loneliness
- unchecked emotion
- desire for immediate closure
- discomfort with uncertainty
- fear of appearing difficult
- overconfidence
- inconsistency

These weaken a person more than any attacker's ingenuity.

REAL WORLD SCENARIO: THE DIFFERENT PERSON

Two people receive the same message:

"Urgent. Need your help immediately."

One acts quickly. The other pauses, evaluates, verifies, and upholds their boundary.

The difference is not the threat. It is the operator.

The G3TI Defensive Identity Shift

From:

- reactive
- available
- pressure sensitive

To:

- emotionally steerable
- assumption driven
- observant
- disciplined
- verification based
- emotionally governed
- process anchored

This is not only safer. It is stronger.

The Highest Form of Personal Defense

The highest form of defense is not suspicion but stability. **A stable person:**

- does not collapse under urgency
- does not chase every disturbance
- does not let fear decide
- does not let shame isolate

- does not let anger direct behavior
- does not let false trust bypass discernment

Stability makes you hard to move, and in a manipulative age, that is power.

The Hardest Truth in This Chapter

No framework will protect a person who refuses to be disciplined.

The greatest threat is rarely a lack of knowledge. It is the failure to apply knowledge when the moment becomes uncomfortable.

Personal defensive intelligence is not only learned; it is practiced, until it becomes who you are.

Final Strategic Shift

The goal is not merely to avoid the next deceptive situation. The goal is to become the kind of person whose life makes manipulation:

- harder to enter
- harder to escalate
- harder to isolate
- harder to pressure
- harder to complete That is the deeper victory.

Final Close

The strongest people in this era will not be those who know the most buzzwords or own the most devices.

They will be the people who have built a disciplined inner system, one that protects attention, regulates emotion, verifies trust, reduces exposure, and refuses to let pressure dictate decisions.

Defensive intelligence is not just about surviving a manipulative world. It is about becoming so grounded, so clear, and so structured that the world cannot easily pull you out of your own self.

Once that happens, you do not merely avoid manipulation. You rise above it.

CHAPTER 17 TAKEAWAY

Personal defensive intelligence is not a tactic. It is a disciplined way of life that keeps your judgment intact under pressure.

CHAPTER 18

The Future Of Trust

How Families, Institutions, Communities, and Nations Must Rebuild Truth, Verification, and Human Judgment in the Age of Synthetic Deception

There was a time when trust came more naturally.

A voice carried weight. A photograph meant something. A video could settle an argument. A familiar name lowered suspicion. A message from a leader could be trusted. A known institution could be relied on without constant doubt.

That world is changing.

Not because trust is no longer needed, but because the signals people once used to gauge trust are being altered, copied, weaponized, and scaled in ways society was never prepared for.

We now live in a world where:

- A child's voice can be cloned.
- A leader's message can be imitated.
- A stranger can build artificial intimacy

- A video can be persuasive without being authentic.
- A false narrative can outpace a verified one
- An institution can be pressured by a lie before the truth arrives
- A family can act on fear before facts emerge
- A community can respond to a manipulated signal as though it were reality

This is not a minor transition. It is a civilizational one.

When appearance, sound, familiarity, and emotional certainty are no longer enough, society must answer a deeper question:

What will trust be based on now?

The Trust Crisis of This Era

Every era has its defining vulnerability.

This era is shaped by a different kind of threat:

the erosion of confidence in what is real, who is real, and when something should be believed.

This is a trust crisis, and trust crises do not stay contained. They affect:

- relationships
- commerce
- governance
- justice

- education
- faith communities
- public safety
- leadership
- family systems
- mental health
- social cohesion
- national stability

If people trust too easily, they become easy prey. If they trust nothing, society fractures into paranoia and exhaustion.

The answer cannot be naïve trust, nor can it be total suspicion. It must be something stronger:

disciplined trust

Why Trust Cannot Be Abandoned

In times like these, cynicism can feel tempting.

"Nobody is real." "Nothing can be trusted." "Everything is suspect." "Every institution is" compromised."

Understandable, but unsustainable. Human life cannot function without trust.

Families need trust. Communities need trust. Markets need trust. Justice systems need trust. Leadership needs trust. Emergency response needs trust. Children need trust. Recovery needs trust. Civil society needs trust. Without trust, only fragmentation remains.

So the answer is not the destruction of trust. It is the reconstruction of trust on stronger foundations.

We are not trying to create a fearful society. We are trying to create a wise one instead.

The Old Trust Model Is Breaking

For generations, trust was built on signals like:

- tone of voice
- appearance
- familiarity
- institutional branding
- position or title
- confidence
- emotional connection
- social proof
- surface consistency

These signals still matter, but they are no longer sufficient.

What once served as evidence now serves only as an indicator.

Indicators alone are not enough when deception is fast, personalized, and convincing.

We are moving from a world where trust was built on presentation to one where trust must be built on verification.

That shift affects everything.

The New Foundation: Verified Trust

The future of trust belongs to those who understand. Trust must not disappear. It must mature.

Verified trust means:

- Identity is not assumed from appearance.
- Urgent claims are not accepted without confirmation.
- Authority is respected but not blindly obeyed
- Emotional closeness does not eliminate discernment.
- Institutions do not treat speed as proof.
- Communities do not share what they have not tested.
- Families do not act on fear without checking.
- Leaders do not expect compliance without accountability.
- Systems support truth, not merely movement This is the future: Not less trust: BETTER TRUST.

The Future of Trust Starts in the Individual

The future does not begin in policy documents. It begins with the person. A future of stronger trust begins with individuals who:

- pause
- verify
- regulate emotion
- resist urgency
- evaluate intent

- preserve evidence
- manage exposure
- refuse manipulation without becoming hardened

If people are unstable, institutions become unstable. If people are undisciplined, systems become vulnerable. If people give in to pressure, manipulation scales.

The future of trust is not only technological but human.

The Future of Trust in Families

Families are the first system of trust most people ever know.

In the future, families must:

- normalize verification without treating it as disrespectful
- teach children that digital familiarity is not proof of anything
- teach elders that checking is a sign of wisdom, not weakness
- build communication protocols for emergencies
- reduce oversharing
- create a home culture where uncomfortable situations can be reported without fear of retaliation

A strong family will say:

We trust each other deeply, and because of that trust, we verify what needs to be verified.

The Future of Trust in Communities and Faith Spaces

Communities shape identity, meaning, and collective action.

But undisciplined trust can turn communities into vehicles for manipulation.

The future requires:

- compassion without gullibility
- unity without blind conformity
- urgency without panic
- leadership without unaccountable authority
- generosity without procedural weakness
- care without chaos Communities must teach:
- verify before forwarding
- confirm before funding
- pause before panicking
- check the source before spreading
- protect the vulnerable by slowing the group down Openhearted and sharp-minded, that is the future community.

The Future of Trust in Business

Markets run on trust.

Businesses must stop relying on informal assumptions and begin embed-ding verification in daily operations.

Outdated habits must end:

- acting on title alone

- assuming internal messages are authentic
- letting urgency override approval
- treating verification as delay
- relying on intuition instead of protocol **The trusted business of the future will be:**
- efficient
- verifiable
- disciplined

Nothing destroys confidence faster than a polished organization that cannot distinguish genuine instructions from manipulated ones.

The Future of Trust in Government and Public Institutions

Government depends on legitimacy. Legitimacy depends on confidence. Confidence depends on trust.

But institutions now operate in an environment where

- authority can be impersonated
- communications can be faked
- rumors can outpace clarification
- public reaction can be shaped by manipulated signals **Public institutions must:**
- communicate clearly
- avoid amplifying unverified information
- model calm verification
- build public education around influence threats

- preserve legitimacy through discipline, not panic

Citizens will trust institutions not because they demand trust, but because they demonstrate disciplined clarity under pressure.

The Future of Trust in Law, Justice, and Evidence

What happens when the evidence itself becomes contestable?

If video, audio, screenshots, and identities can be forged, justice systems must become more disciplined.

Truth does not disappear, but it must be established with greater care. This raises the value of:

- corroboration
- metadata
- sequence analysis
- independent validation
- disciplined evidence handling
- context based assessment
- trained skepticism without nihilism

These themes matter far beyond individual safety. They shape the future of legal and social order.

The Future of Trust in Leadership

Leadership will require a different kind of strength. Not charisma. Not speed. Not force.

But disciplined credibility. The leader of the future:

- does not spread unverified information
- does not use urgency carelessly
- does not bypass process
- does not reward panic
- does not weaponize authority
- does not punish caution
- does not confuse decisiveness with recklessness

People mirror leadership. If leaders panic, groups panic. If leaders verify, groups verify.

The future of trust depends on leaders who safeguard clarity.

The National Dimension: Trust as a Strategic Asset

A society that cannot tell what is real becomes easy to manipulate on a large scale.

This affects:

- elections
- public safety
- market confidence
- national security
- emergency response
- institutional legitimacy
- crisis stability
- public order

Trust must be treated as a strategic asset. Nations that remain stable will be those that:

- train citizens in discernment
- harden institutions around verification
- protect communities from narrative manipulation
- preserve evidence standards
- normalize disciplined trust

This is not just about avoiding scams. It is about whether society can function when deception is easy to manufacture.

The Great Danger: Two Bad Futures

If society fails to respond wisely, it may drift toward one of two undesirable futures.

Bad Future One: Mass Gullibility

People trust appearance and urgency without lowering their standards. Bad Future Two: Total Cynicism People trust no one and nothing. Neither future is acceptable.

The answer is a third path:

wise, disciplined, structured trust

What the Next Generation Must Be Taught

The next generation must learn:

- how to pause under pressure
- how to question urgency respectfully

- how to verify identity and source
- how to distinguish confidence from credibility
- how to protect attention
- how to identify emotional manipulation
- how to preserve evidence
- how to report without shame
- how to use technology without surrendering judgment
- how to trust wisely instead of blindly This must become common civic knowledge.

The Moral Dimension of This Era

If deception becomes easy, what kind of people will we be?

Will we exploit confusion or protect clarity? Spread emotion or hold in-formation until tested? Demand trust or earn it? Panic or verify? React or think?

The synthetic era will test character as much as it does technology.

Tools amplify the person who uses them. If discipline is weak, damage multiplies. If discipline is strong, truth still has a future.

The Future of Trust Is Not Automatic

It must be chosen.

People, families, institutions, and communities must commit:

- We will not move on urgency alone.
- We will not spread what is unverified.

- We will not let emotion outrun truth.
- We will not let title replace proof.
- We will not abandon trust.
- We will rebuild it with stronger foundations.

The G3TI Trust Reconstruction Doctrine

T.R.U.T.H. Model

T — Test before trusting: Nothing moves forward on appearance alone.

R — Rebuild through verification: Trust strengthens when it withstands scrutiny.

U — Understand pressure dynamics: Urgency, fear, authority, and emotion are influence factors, not proof.

T — Train at every level: individuals, families, leaders, institutions, and communities.

H — Hold the line: Do not lower standards because the moment feels intense.

This is how trust survives the future.

What This Book Has Really Been Building

This book has been building a worldview, a way of operating, in which:

- The mind is protected.
- The emotions are governed.

- Evidence matters
- Verification becomes a habit
- Communities become harder to manipulate
- Institutions become harder to exploit.
- Public safety evolves
- Trust is preserved through discipline.
- People stop reacting under engineered pressure and start controlling their response

This is not merely a defensive system. It is a stabilizing force.

The Final National and Human Challenge

The great question of the coming years is not whether AI becomes more powerful.

It is whether human beings remain wise enough to govern themselves in a world where falsehood can masquerade as truth.

The real battle is:

- truth versus speed
- clarity versus confusion
- discipline versus impulse
- verified trust versus engineered belief
- mature judgment versus manipulated reaction The winning side will shape the future.

Final Close

The future of trust will not be secured by nostalgia or fear. It will be secured by people who understand that truth still matters, that trust is still possible, and that human judgment, when disciplined and anchored in verification, can still rise above deception.

We cannot return to a simpler age when appearance was enough. But we can build a stronger age in which trust is not abandoned, only refined. Ultimately, the greatest defense in the age of synthetic deception will not be found in code alone or in policy alone.

It will be found among people who refuse to let urgency outrun truth, emotion outrun judgment, or appearance outrun verification.

If we become those people, individually, collectively, and nationally, then trust does not die in this era.

It evolves. It matures. It survives. And so do we.

CHAPTER 18 TAKEAWAY

The future belongs to people and institutions that preserve trust through verification, discipline, and sound judgment.

May God bless you, strengthen you, and protect you from every form of deception.

"But the Lord is faithful, and He will strengthen you and protect you from the evil one."

— 2 Thessalonians 3:3

www.ingramcontent.com/pod-product-compliance
Lightning Source LLC
LaVergne TN
LVHW100308210926
848480LV00003B/1056
* 9 7 9 8 9 9 6 3 4 8 4 0 4 *